To our dear friends, Susan and Michael,
in honour of our friends John Brown –

With love
Chris and David

JOHN C BROWN is Professor Emeritus at the University of Glasgow and has been Astronomer Royal for Scotland (ARfS) since 1995. In astronomy research he has published around 300 papers and won the 2012 Royal Astronomical Society Gold Medal. A keen teacher and semi-professional artist and magician, he is wizard at lucid explanation and illustration of amazing science ideas. As ARfS he gives numerous astronomy talks to all audiences and ages and collaborates widely on creative arts projects. His 2016 OBE was for 'services to promotion of astronomy and science education'.

RAB WILSON is one of Scotland's most accomplished poets. His poetry in Scots is distinguished by its oral quality and by its technical assurance. The author of a number of highly praised volumes of poetry, he is also a Burns scholar. He has worked with the artist Calum Colvin on a book of responses to Burns and he is the Scriever in Residence for the National Trust for Scotland based at the Robert Burns Birthplace Museum in Ayr.

(Figure 1.0.1) Horsehead Nebula region – Orion molecular cloud complex
(Graham Malcolm, Kilbarchan)

Oor Big Braw Cosmos
A Cocktail of Cosmic Science, Imagery and Poetry

JOHN C BROWN and RAB WILSON

Luath Press Limited

EDINBURGH

www.luath.co.uk

Dedicated to the memory of Scots genius James Clerk Maxwell
and his unparalleled insights into the workings of oor big braw cosmos

First published 2019

ISBN: 978-1-913025-05-2

The paper used in this book is recyclable. It is made
from low chlorine pulps produced in a low energy,
low emission manner from renewable forests.

Printed and bound by CPI Antony Rowe, Chippenham

Typeset in 10.5 point Sabon by Main Point Books, Edinburgh

Contents

Foreword

by Regius Chair Professor Andy Lawrence FRSE (Edinburgh)

I WAS VERY pleased when asked to write a Foreword for this wonderful new book. The idea of combining science, images and poetry in a coherent whole was very intriguing. I had been hearing hints of its progress for some time, and was looking forward to seeing what strange beast would emerge. I was not disappointed.

I love science, and I love art, and so I love to see them collide and combine. In both spheres, my personal preference is for the concrete and direct, rather than the abstract or elusive. Being concrete doesn't mean you have to be prosaic. Contemplating what we have learned about the cosmos produces in most people an awestruck and emotional response. Likewise, in well crafted poetry, it feels like each word plucks a string in your head, and each string is connected to seven other strings. In all these ways, this book is satisfying. The science is straightforwardly explained, and links directly to the beautiful images; and the poetry, while lively and robust in a man-on-the-street style, has all those little resonances that set deep things off in your mind. The whole thing links together in a delightful way.

As John Brown himself describes, we seem to have been living through an epoch where art and science have been on divergent paths, or at least require very different modes of thinking, and so are carried out by separate breeds of people – but it was not always thus. Da Vinci of course somehow managed to straddle both worlds like a god, but that's hard to live up to. A better example perhaps, described in this book, is the career of Victorian astronomer Charles Piazzi Smyth. He was an innovative and important scientist, but he also sketched and painted throughout his life, and was a pioneer of photography. He was not the best artist of Victorian Britain, but he was a seriously creative one – his drawings have life, interest, insight and beauty.

Much though the modern scientist might dream of being da Vinci or even Piazzi Smyth, it's a very hard trick to pull off in modern times, because being a professional working scientist is so hard – it takes years of rigorous training and practice in a narrow specialisation. Likewise, being a serious poet requires years of experiment, thinking, and learning how to polish and arrange the words just so. Meanwhile, however, the scientist does not lose their desire for artistic expression, nor does the poet lose their desire to understand what makes the natural world tick. So, the answer is *collaboration*. I am sure that John Brown and Rab Wilson will have had a marvellous and fruitful time in putting this creation together, sparking ideas and slotting them together. Two heads are better than one, as this lovely book shows.

Andy Lawrence
January 2019

Foreword

by Visiting Professor Dame Jocelyn Bell Burnell FRS, FRSE (Oxford)

ASTRONOMY IS VERY visual. All of us (who have sight) will have seen at least the Sun and the Moon, and probably also a star or planet or two, which arguably makes astronomy the most accessible of the physical sciences. Such sights are often, in addition, evocative, stirring, wondrous. And, when we learn of the distances involved and of the workings of the Universe, as we do in this book, how much more so!

So perhaps it is surprising that there is not more poetry, art, music, sculpture and literature addressing this realm. For a number of years, I have been 'collecting' poetry with an astronomical theme and it turns out that there is quite a lot of it, though it is rather hidden (it rarely features in anthologies, for example). The main exception is poetry about (or addressed to) the Moon – and I note that, similarly, there are a fair number of 'Claire de Lune' musical compositions. If spectacular comet apparitions were more frequent, there might have been more on that theme, especially as poems often address the transient nature of life.

The literary world is by no means devoid of the scientifically minded. Thomas Hardy was, for example, an amateur astronomer and it shows – he gave a brilliant description of the scale of the Universe in *Two in a Tower*. More recently, Rebecca Elson was a very able professional astronomer as well as a wonderful poet.

This volume, with its use of poetry in Scots to give verbal colour to cosmic beauty and wonder, adds an interesting new dimension to the subject and will also reach out to people who otherwise might not venture into such an area. It is good also to see Scots promoted in this way – recognising and honouring all forms of diversity is important.

I note, wryly, that when I do a talk on an astronomical topic my audience is predominantly male, but when I do a talk about astronomy and poetry there are lots of women present! Hopefully, the interweaving here of science and art, both verbal and visual, will help us transcend not only the culture (science/art) divide but also the gender one that seems still to be with us, by showing how different sides of diversity can enrich each other, to the benefit of all.

Jocelyn Bell Burnell
January 2019

Introduction

IF THERE IS ever a contest for the era in which the Universe around us delivered the greatest excitement, a strong contender must be that which (coincidentally!) encompasses the lives of your two authors. One was born soon after the Jodrell Bank Observatory opened, and the other seven months before Gagarin orbited the Earth. Radio astronomy and spaceflight, which blossomed from these opened amazing new windows on the Universe in terms of the range of wavelengths (colours) that could be seen and the ability to *explore* rather than just look at cosmic objects.

Over the last half century or so, since the first manned Moon-landing, the wavelength span of these windows has soared from the 2:1 range of visible light (the range detectable by the human eye) to around 10^{21}:1 for very low frequency radio to the highest frequency gamma-rays. Further, modern technology has added to this non-electromagnetic data from cosmic neutrinos and gravitational waves. Today, we are in the era of multi-messenger astronomy, combining as many as possible of these data sources as witnesses in the cosmic Crime Scene Investigations (CSI), which are the lifeblood of astronomers. Telescope systems in near-Earth orbit are probing everything imaginable and finding many things that were, until recently, unimaginable, such as the thousands of now-known exoplanets (planets around stars other than the Sun) and estimates of billions of habitable Earth-like planets in our Galaxy alone.

In parallel with all this remote telescopic observation of the cosmos, the burgeoning international activity in space probe launches has carried sophisticated robots to examine closely every planet of our Solar System and to land on some of them, as well as on asteroids and comets, in some cases bringing back surface samples. It has been hard to keep up with breakthroughs in space research even during the writing of this book. In the first few weeks of 2019 alone: Japanese mini robot probes have been hopping around on an asteroid looking for a main probe landing site; a Chinese lunar rover has landed on the Moon's far side; and America's New Frontier Pluto probe, now at a distance of 6.4 billion km, has discovered that Asteroid Ultima Thule is likely a contact binary – a pair of objects orbiting the sun together but held in contact solely by their own gravity. This is truly staggering progress.

As yet, we have not found life beyond Earth. However, the discoveries of so many habitable exoplanets, of life in very extreme terrestrial environments, and of so many new facets of biology, have led to a global explosion of interest and research in astrobiology, a subject that has no real data yet but has colossal potential. The last 100 years brought the realisation that almost all of the atomic elements in our bodies were created and made available by stars. It is surely only a matter of a little more time before we understand how those elements are synthesised into the complex molecules and elaborate organic compounds we are finding on comets, moons and planets, then on to how life-defining processes of self-replication work. In short, we now know for sure that we are made from star-stuff – but not yet exactly how the making works.

This era has fortunately also seen a healthy movement away from the groundless but damaging divide between arts and science addressed by CP Snow in his renowned 'Two Cultures' lecture and discussion. (This seems to have been a symptom of our particular era. Earlier history is punctuated with visionaries like Omar Khayyam and Edgar Allan Poe with interests spanning poetry, astronomy and mathematics.) Many of today's festivals, be they science, arts, literature, or music, involve elements mingling two or more of these areas, creating the genre of 'sci-art'. In the UK alone, these include annually the huge two-week Edinburgh SciFest (one of Europe's largest science festivals); Orkney's major SciFest; Jodrell's Bluedot Fest Weekend; and, in 2019, the first Hebridean Dark Skies festival at An Lanntair in February. There is now even a major international body dedicated to cosmic art – the International Association of Astronomical Artists (www.iaaa.org). So, there is no doubt that CP Snow's Two Culture (arts-science) divide is rapidly evaporating. As Jocelyn Bell Burnell points out in her Foreword, there is still a degree of gender divide in science, or in some sciences at any rate – especially physics – but that, too, seems to be on its way out, albeit slowly. Hopefully today's great new level of sci-art activity, and even oor wee book, will help catalyse further gender inclusivity.

There are many recent stars in the sci-art genre. Lynette Cook's diverse space-art has been used in many recent astronomy texts while Ekaterina Smirnova uses ingenious touches like incorporating heavy water in the paint medium for her comet paintings, displayed at many science meetings. Her work can be seen at www.ekaterina-smirnova.com. There are the free-form electronic improvisations by Herschel 36 (Paul Harrison and Stu Brown), which accompany silent film-fest and sci-fest showings of *Wunder der Schöpfung* (1925), the first full-length space documentary. There is the cosmic poetry of the late Rebecca Elson, particularly her anthology *A Responsibility to Awe*. There are cosmic landscaping and installation artworks by the likes of Antony Gormley (*Blind Light*), Charles Jencks (Portrack *Garden of Cosmic Speculation* and *Crawick Multiverse*), Gill Russell (*Sòlas*) and James Turrell (*Skyspace*). More on these is to be found in Chapter 6. On a global level, the guidance of Caltech's Nobel Laureate Kip Thorne on the science content of the blockbuster movie, *Interstellar*, led to breakthroughs in black hole visualisation, and a joint scientific paper on the findings.

Mention must also be made here of the massive role played recently in some sciences by citizen science (CS), whereby armies of enthusiastic members of the general public harness the colossal power of the internet to search and categorise huge databases in ways that automated machine searches cannot and for which professionals simply do not have time. In astronomy, a major CS project within Zooniverse (www.zooniverse.org) is Galaxy Zoo (zoo1.galaxyzoo.org), organised by Professor Chris Lintott of Oxford and the BBC's *Sky at Night* programme. More on this in Chapter 4.

So, it is with enthusiasm and, we hope, timeliness in this 50th anniversary of the first Moon landing, that we jump into this exciting field of blossoming sci-art, with our personal takes on the wonder and beauty of our colossal cosmos. We mostly address the beauty and quirks of the cosmos itself, and of its scientific workings, but also introduce some colourful people who have researched, contemplated and portrayed their work in both science and arts across the aeons. While we are first and foremost a scientist and a poet, we have both long had eclectic interests and involvement across disciplines. We share a passion for riding Glasgow's famous Flying Scot lightweight bicycles, Rab having written a poetry collection about them with music to go with it. John is a semi-pro magician, using his skills to illustrate extreme science like black hole phenomena. He also tries his hand with cosmic graphic art ideas, and works closely with writers, musicians and artists. Here we combine our passions and skills to bring together the forces of imagery (direct and artistic), of scientific thinking, and of poetry to bring to life the beauty of our vast cosmos for as many readers as possible. John had long had the notion of creating such a book, but it only came to be as a result of one of those cosmic collisions in life, in this case meeting Rab at a Drumlanrig Castle dinner hosted by Richard Scott, Duke of Buccleuch, following a sci-art event based around Charles Jencks' Crawick Multiverse.

Some remarks are in order regarding the Scottish slant of this book and its title. These are in part driven by the remarkable Scottish contribution to astronomy over the centuries as described in the opening of Chapter 5. Also, we have chosen to include a lot of the fabulous images being taken today by amateurs even in the high humidity, often cloudy and very unpredictable weather of Scotland and places like it. These are far better in detail (though not in light-gathering) than images from the world's largest telescope (Mount Palomar) in the 1950s and sometimes hard to distinguish even from Hubble Space Telescope results. This is down to modern telescope technology and even more so to computer (image processing) power and concepts. We are very indebted to all those numerous amateur astrophotographers who offered us their work, and regret that we simply could not find space for them all.

On the matter of our choice of Scots as the language for the many new poems presented here by Rab, this is in part as a tribute to Scotland's standing in cosmic studies and in part because Scots is a highly expressive language. This is widely accepted worldwide as a result of our renowned national poet, Robert (Rabbie) Burns, who wrote in Scots about his many passions in and for life. It is, therefore, an ideal language for a book full of passion for the brawness of all things cosmic. Like English, Scots has its roots in the old Germanic languages of the tribes that invaded Britain after the Roman occupation. Like Old English it can be fairly readily understood after a few readings. In fact, it is easier for all, not just Scots, to read than Chaucerian and even Shakespearean English, as is shown by the worldwide popularity of Burns' poetry. Here we mostly use relatively straightforward Scots which after a read or two you should find as easy to grasp as the scientific content. Readers encountering any difficulty with the Lallans can dip into *Dictionary of the Scots Language* (*Dictionar o the Scots Leid*) at www.dsl.ac.uk, which is an utter joy to read in itself. Similarly, if anyone gets a little stuck with science terminology, they should turn in the first instance to www.wikipedia.org/wiki/Glossary_of_astronomy. Web resources like these are now a far more versatile and helpful supplement to a book like this than appending a glossary.

So, here we are in print at last and, despite lots of hard graft, it has involved shedloads of both fun and learning. We hope all our readers get some insight, pleasure, and chuckles from these pages, and we would urge them to punctuate their reading with at least occasional forays outside to admire oor big braw cosmos for themselves. This could be to view some shower meteors, the stars or, of a winter evening, the fuzzy patch in the middle of Orion's sword, which is the beautiful Great Nebula of Orion (Figure 1.0.2) – a star and element factory producing the star stuff that makes us all (see Section 3.4).

John C Brown and Rab Wilson
April 2019

Figure 1.0.2 Orion Nebula on a Christmas card showing a stellar nativity scene with two satellite tracks
(Douglas Cooper, Doune)

Acknowledgements

WE WISH TO thank the many folk who have lent invaluable support of diverse kinds, not least our families and friends for their tolerance and support through this major undertaking. Travel funding from the Royal Society of Edinburgh (RSE) Cormack Bequest, from the Society of Photographic Instrumentation Engineers (SPIE), and from the University of Glasgow was invaluable in our book planning and progress meetings. Proofreading and comments on content at draft stages by Margaret Brown, Andrew Conway, Dugald Cameron, Arthur Jones, Lorna McCalman and Keri Simpson kept us on the straight and narrow while sustaining our enthusiasm, while all Luath Press staff provided much editorial help and many suggestions in the design and production stages.

Turning next to contributors to the book's content, the Forewords by Professor Dame Jocelyn Bell Burnell (Oxford) and Professor Andy Lawrence (Edinburgh) are delightful bonuses to our aims, given Jocelyn's great passion for cosmic poetry and Andy's arts interests, as well as their eminence in astronomy research (Jocelyn being the famed discoverer of pulsars). Above all, however, we have to thank the many and varied sources of the extensive imagery used in this book to convey the brawness of our cosmos and its workings to our readers. Free web search facilities offered by many sources like Wikipedia and Wikimedia Commons, and within more specific bodies like NASA and ESA greatly simplified the finding of many suitable images. However, that in itself would not have enabled this book had it not been for the vast generosity of so many organisations, individuals and friends in allowing to use their material free of charge, whether as a matter of general policy (eg NASA), or because they felt this book would nurture our mutual belief in the importance and value of cross-cultural inspiration and education. The sheer number of sources involved is too lengthy to list in this introduction, but we credit them all in the relevant text and Figure captions. That said, special thanks go to all contributing amateur astrophotogtaphers, and to those professionals (like Lynette Cooke, Charles Tait and Martin Shields) who permitted us to reproduce copies of their superb creative works, on the sale of which their livelihood largely depends.

We also wish to thank the National Trust for Scotland and the Robert Burns Birthplace Museum – especially Learning Manager Chris Waddell – for its championing of the Scots Language and the practical support and cultural encouragement provided for Rab Wilson's poetic input to the book.

The extracts in the Appendix from 'We Astronomers' are quoted from the collection *A Responsibility to Awe* by Rebecca Elson (1960–99), edited by A Berkeley, A di Cintio, and B O'Donoghue, published by Carcanet Press and reproduced here with their kind permission (© Estate of Rebecca Elson).

Every effort has been made to ensure that appropriate credits and copyrights are indicated for all images and quotations used in this book. We apologise for any inadvertent errors or omissions; however, should you come across any, please notify john.brown@glasgow.ac.uk. Permissions of use are not transferable and apply solely to the authors and to this publication. These permissions in no way imply endorsement of content or views expressed here.

Prologue: The Makar

Aathing hus its plan;
The Makar taks his compass,
An measuirs aathing oot;
Cubits lowp tae licht years,
Frae Alpha til Omega.
Thon void athoot form,
Thae heivins that inhaud it,
Licht o daw an lesser licht o nicht,
The lift fill't wi undeemous starns…
At aince unkent an yet, fameeliar.
Einstein's eldritch mystery,
Bides aye its reddin up.
But syne;
A muckle bang!

Figure 1.0.3 Big Bang, 2008
(Cédric Sorrell, commons.wikimedia.org/wiki/File:Big_bang.jpg)

CHAPTER I

Let's Get Started

Think o a nummer,
The biggest nummer evir;
It's faur ower wee.

1.1 Big Numbers and Just How Big *Is* the Universe?

AS DOUGLAS ADAMS says in *The Hitchhiker's Guide to the Galaxy*, 'you may think it's a long way down the road to the chemist's, but that's just peanuts to space.'

In this book, we aim to help those with a phobia of large numbers to get their heads around the astronomically large values that arise when talking about such things as numbers of stars and cosmic distances. We do so in part by using analogies and comparisons with more familiar everyday items and situations. First of all, however, we explain the scientific notation commonly used for large and small numbers for those not familiar with it, since we use it in some parts of this book.

When dealing with very large quantities, instead of writing out long strings of digits such as ten thousand billion billion – 10,000,000,000,000,000,000,000 (the approximate number of stars in the visible Universe) – we can denote this by 10^{22}, where 22 (the exponent) indicates the number of zeroes. Thus, for example, one hundred = 100 = 10^2; one thousand = 1,000 = 10^3; and one million = 1,000,000 = 10^6, while for 5 million, for example, we write 5×10^6. For numbers less than one, the scientific convention is to use a negative exponent. For example, one tenth = 1/10 = 0.1 = 10^{-1}; one thousandth = 1/1000 = 10^{-3}. (In astronomy, as well as very large numbers, we encounter very small ones, such as times in the very early Universe.) In this book, depending on the context, we use a mix of these notations, using the scientific 10^N notation when wishing solely to state a numerical value, but words when we want really to convey the vastness of something – ie ten thousand billion billion stars grabs the attention more than 10^{22}.

In describing physical quantities (like mass, length, time, energy etc), as opposed to pure numbers, astronomers tend to use a mixture of physical units, to the chagrin of physicists who try to stick to the Standard International system based on the metre,

Sky Bridge
JC BROWN 2012

Figure 1.1.1 Sky Bridge, 2012
(JC Brown, Glasgow)

kilogram and second (mks). We do so partly for historical reasons and partly because of the huge range of values of quantities occurring in astronomy. We will introduce and explain such special units (like Astronomical Unit AU and Light Year LY) as and when we first use them here. For now, suffice to say that it is easier and clearer to say a star is about 10 LY away rather than about 10^{14} km, and to say its mass is 10 Solar masses rather than 2×10^{31} kg. For more everyday distances and masses, we will mostly use metric measurements like kilometre, metre or centimetre (km, m, cm), and kilogram or gram (kg, g). For those Americans and some Brits still using Imperial units, we occasionally mention Imperial equivalents.

Turning now to the vastness of numbers that describe things in *Oor Big Braw Cosmos*, these scare some people off since they say they can't get their heads around them. To put this in perspective, let us consider the oft-cited statement, 'There are more stars in the Universe (over 10 thousand billion billion = 10^{22}) than there are grains of sand on all the beaches on Earth'.

Mair starns in heivin,
Than saund oan aa the beaches –
Ken whit ah'm sayin?!

However, the fact is that there are more atomic nucleons than that in one fingernail. To confirm this, note that the mass of all chemical species of atoms lies almost entirely in their central nuclei, made up of protons and neutrons (nucleons), each of which has a mass of about 1.67×10^{-27} kg = 1.67×10^{-24} g (the atomic electrons are almost 2,000 times lighter). A fingernail weighs around 1 g, so contains $1g/1.67 \times 10^{-24}g = 6 \times 10^{23}$ nucleons or 60 times the number of stars in the Universe. In fact, numbers relating to the microscopic world inside us are mostly even more amazing than the cosmic ones, while those describing life, the brain, consciousness and complexity are gobsmackingly larger

Figure 1.1.2 Dense starfield NGC4833
(NASA/ESA HST www.spacetelescope.org/images/potw1631a/)

still. For example, a simple pack of playing cards comprises just 52 distinct cards, but the number of distinct sequences in which they can be arranged is around 10^{68}. These are the kinds of numbers involved when considering how many tunes one can make from a piano keyboard, or how many words from a 26-letter alphabet.

Moving on from the huge values that arise from counting numbers of things, let us have a look at the physically enormous size of oor big braw cosmos by taking a trip across it, starting with the Solar System. Driving at about 50 mph non-stop, we could cover roughly 1,200 miles a day. Thus, driving to the Moon would take about 6.5 months; to the Sun, it would take about 300 years; and to Pluto, it would take about 10,000 years. Since none of us has that long, let's crank up our speed 1,000 times to 50,000 mph. Then we can get to the Moon in about 5 hours, to the Sun in about 3 months and to Pluto in about 10 years – still a bit slow for human lifetimes. So now let's push things to near Einstein's ultimate speed limit – the speed of light, or 300,000 km/sec – a factor of 12,500. Then, in times observed from Earth, we could reach the Moon in roughly 1.3 seconds, the Sun in roughly 8 minutes, and Pluto in roughly 8 hours. These distances are then termed 1.3 Light Sec, 8 Light Min, and 8 Light Hr. So, if two student pals planned different outings, one to the pub and one across space at near the speed of light, one would pass the Moon by the time her pal had lifted her first drink, pass the Sun by the time her pal had downed it, and pass Pluto when her pal was heading home at closing time.

Now let's look beyond this already huge Solar System of planets to the distances of the nearest stars, first measured in 1836 independently by Friedrich Bessel, Friedrich van Struve and Thomas Henderson, Scotland's 1st Astronomer Royal (of whom there will be more later in Section 5.5). The nearest star, Proximal Centauri is 4.2 LY away, where 1 LY = the distance travelled at speed c in 1 year = approximately 10^{13} km, or about 200,000 times further away than the Sun. Thus, by the time our light speed space-tripping student reaches even the nearest star, her student drinking pal will have enjoyed over four years of pub visits and, if she fitted in enough work too, should have graduated. The reason is simply that the distance to even the nearest stars is around 10,000 times bigger than that across the entire Solar System, thought for over 2 millennia to be the whole Universe. This all goes to show that even nearby interstellar travel is a challenge to humans simply in terms of taking up a good chunk of a lifetime. However, it is much more challenging than that when one considers the fuel energy bill involved. It is easy to show that to accelerate an average person up to even half the speed of light would require around the entire energy consumed by the whole of humanity in one year. So, interstellar travel is not a gift anyone can expect to get from Santa.

These are utterly gobsmacking facts arising from the distances to the stars. However, they pale into insignificance when compared with what astronomers have progressively discovered over the two centuries since Henderson *et al.* concerning just how far the Universe stretches beyond nearby stars. The distances to nearby stars are measured using triangulation (parallax), just as is done by surveyors and builders using movable theodolites (which accurately measure the directions of features). You can easily see the idea of parallax distance measurement yourself by holding out a finger in front of your face and looking at it alternately with your left eye and then your right eye. Relative to background objects, the finger seems to jump back and forth, but the further you hold it away from your face, the smaller the parallactic shift it shows. That shift can measure distance. In the case of the much longer baseline of the Earth's motion around its orbit, the shift in viewpoint is the Earth's orbital radius (150 million km). This motion causes the direction of nearby stars to move back and forth against faint distant background ones by about one arc-second – about the angular size of a penny seen from 2 km away. For remote stars, these angles become immeasurably small and we have to resort to indirect methods such as the Standard Candle technique of calculating distances by comparing them to the apparent faintness of distant objects whose true brightness we know by other means. The idea is based on the fact that how bright something appears depends on how luminous the object itself is and on how far away it is. (Likewise, how large something looks depends on its intrinsic size and on its distance – as in the famous sketch where Father Ted explains to Father Dougal the apparent sizes of real and toy cows.)

A striking illustration of the issue of apparent brightness is shown in Figure 1.1.3, an image taken on 4 April 2013 by Douglas Cooper of Comet PanSTARRS crossing the sky near the Andromeda Galaxy M31 (Section 4.1). Superficially, these two objects look rather similar – fuzzy and with about the same angular size and same apparent

Figure 1.1.3 Comet PanSTARRS, M31 and meteor, April 2013
(Douglas Cooper, Doune)

brightness. In reality, M31 is physically around 10^{11} times larger and 10^{22} times more luminous but is 10^{11} times further away (2.5 million LY versus around 10 Light Min). In fact, this interesting image contains objects at not just these two distances but at *four* very different distances scales. In addition to the comet at tens of Light Min and Andromeda at 2.5 Million LY, there are field stars at distances of tens to hundreds of LY and, by chance, there is also the transient streak of a meteor/shooting star (Section 2.1) burning up in our atmosphere around 0.0001 Light Sec away from us.

Their different distances also affect the motion they exhibit. Meteors, comets and stars mostly have speeds in space in the range of tens of km/sec, while it's more like hundreds of km/sec for galaxies. Meteors in our around ten km-thick atmosphere flash by us in fractions of a second, while comets in the Solar System typically move across the sky over periods of weeks to months. On the other hand, even the nearest stars are over 100,000 times farther away than the Sun, so they take many years to show any perceptible motion across the sky (hence the ancient term Fixed Stars). Being millions of times further away still, the motion of galaxies across the sky will take many aeons to be perceptible. Finally, we note the astonishing spread in the typical masses of these four classes of naked eye objects in Figure 1.1.3: meteor particle $\sim 10^{-3}$ kg; comet $\sim 10^{15}$ kg; star $\sim 10^{30}$ kg; and galaxy $\sim 10^{41}$ kg.

The crucial historic example of a Standard Candle is that discovered in 1912 by Henrietta Leavitt (Harvard): for pulsating Cepheid stars, the pulsation period is uniquely related to the intrinsic brightness (Sections 4.1–4.2). This allowed measurements of the vast interstellar distances across our Milky Way, and enabled Hubble and others to establish that the Andromeda Nebula M31 is a separate galaxy lying far outside our Milky Way Galaxy but the nearest large galaxy neighbour to us. They also discovered that, the more remote galaxies are, the faster they are receding from us as measured by Doppler shift. (This is the compression or stretching of wavelengths due to approach or recession of the source, like the rise and fall in pitch of sound waves from a passing police car siren as its approaches then recedes.) This expansion law was discovered by Slipher but refined later by Hubble and henceforth unfairly credited by the world almost solely to him. Hubble's Law for our expanding Universe has enabled measurement of even greater galactic distances by means of Doppler reddening measurements (more on this in Section 4.1).

Complex as all this may seem, the bottom line results are that the Universe is vastly bigger than even the huge nearby stellar distances of a few LY. In brief, the distance across the Milky Way is around 100,000 LY; to Andromeda M31 is around 2 Million LY; and across the entire visible Universe is the colossal distance of over 10 billion LY. In terms of travel, one surreal way to visualise this is to compare a trip across the whole

Universe travelling at half the speed of light with a train journey from Glasgow to London. In order to make the trips take the same length of time, the speed of the train would have to be about 1 mm per century, making even Britain's trains seem fast.

This vastness also puts the seemingly gigantic number of stars in the Universe in a new perspective. Using the huge number that we started with, there are 10^{22} stars. Spread across the colossal size of the Universe, even this vast number of stars leaves it feeling like a very empty place. The Universe is about 10^{23} km across. Spreading out all the stars uniformly would space them on average 5×10^{15} km (500 LY) apart. This is so empty that, if one flattened a spherical Universe into a uniform disk with the same diameter but with the density of water, it would be about the thickness of a penny (Figure 1.1.4) – a compression factor of around 10^{29}. These figures are based solely on the mass of stars but, even if we add in the roughly comparable mass of gas and dust visible by their emission or absorption and the roughly ten times larger mass of dark matter not visible, that disk would only be the thickness of a sugar cube.

Figure 1.1.4 Cosmos flattened to a thin disk
(JC Brown, Glasgow)

Thair are three atoms
Per cubic metre o space!
Nae sae empty, eh?!

The physical impossibility (Einstein) of reaching light speed c (because it needs infinite energy) and the financial/political impossibility of reaching any significant fraction of c mean we will never travel much, if any, beyond nearby stars, unless we find new science that enables it, such as:

- Figure out how to make and operate whatever a Time Lord's Tardis might be: Figure 1.1.5.
- Find or create wormholes (eg via black holes) from one part of space-time to another through space-time, such as are shown in the movie *Interstellar,* and somehow avoid them shredding us – sometimes called spaghettification (see Figure 1.1.6). The red arrow represents the normal long path through ordinary space while the shorter green arrow represents a wormhole shortcut cutting through another dimension.
- Create a warp drive such as powers Star Trek's USS Enterprise by shrinking the

space-time gap between us and remote objects, without either having to move locally. An example to demonstrate this is to warp a balloon to bring together two spots on its surface, without either moving across the local balloon surface.

The only snag with all of these is likely to be how much energy is needed to make and operate a wormhole, warped space-time, or a Tardis. Rather a colossal amount I think, and much more than a wee stroll to the Moon or Mars like Elon Musk has in mind.

However, in the safe SciFi world populated by characters like American Captain Kirks and middle-class British Doctor Whos, such practical issues are of minor importance. Perhaps our first female Doctor will tell us the trick though – gender and accent aside, she seems to be from the same mould as her predecessors.

Figure 1.1.5 Tardis Coffee,
Glasgow West
(JC Brown, Glasgow)

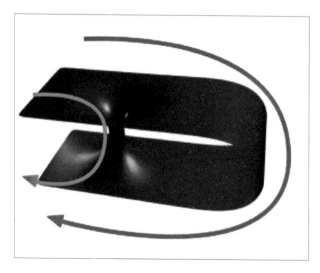

Figure 1.1.6 2-D wormhole analogy
(PANZI, commons.wikimedia.org/wiki/File:Wormhole-demo.png)

Doactir Wha?

In a Cosmos o endless possibility,
Whaur aathing can heppen,
An onythin ye micht jalouse micht be,
Nae leemit til its myriad o life-forms,
The eildritch Doactir chainges yet agane,
Transmogrifies intil...

A white, middle-cless, male British stereotype.

Ye'd think that wi the hale o Time tae play wi,
The hale o Space; frae Gallifrey til Greenock...
Wi beings faur ayont oor puir ingyne,
They micht dae better?

Syne then it heppent!

As Daith's gleg gully raxed fir him yet agane,
Fate's fell Rob Sorbie pierced him tae the hairt,
An he becam...
A wumman!

White, middle-cless, an British.

Figure 1.2.1 Paolozzi Newton sculpture, London
(Andrew Dunne, https://commons.wikimedia.org/wiki/File:PaolozziNewton1995.jpg)

1.2 Space, Gravity, Orbits and the Vacuum

In this section, we have a look at aspects of space and clarify the nature of (outer) space, gravity, orbits and the vacuum. In particular, we correct the common fallacy that there is no gravity in space, though astronauts nevertheless feel weightless. We also explain why space is a (near) vacuum.

For starters, what *is* gravity? Isaac Newton and his less well-known arch-rival, Robert Hooke, were the first to explore the concept of a universal gravitational force, or action-at-a-distance – between bodies. This was devised to explain and unify such diverse events as apples falling from trees and planets falling around (orbiting) the Sun or the Moon around the Earth. Newton's famous Universal Law of Gravity states that this force scales with the product of the masses of the two bodies divided by the square of their distance apart so that, even though the force falls at larger distances, it never goes to zero until their separation is infinite. So, there is nowhere in space where gravity is quite zero. Applying this law to Low Earth Orbiting (LEO) satellites (a few hundred km altitude) shows that Earth's gravity there (and your weight) are under 10 per cent less than on the ground.

So why do we *feel* weightless in space?

In everyday settings, we are standing, sitting or walking on the Earth with the pressure of the solid ground or chair on our feet or bottom balancing gravity. Thus, we neither fall nor rise though we feel heavy because that pressure force is balancing gravity.

By contrast, when we are falling freely – as a lift starts downward, or we jump into a pool, or our plane falls in turbulent air, or we go sky-diving – we have no ground contact so experience no pressure force on our body. In fact, we feel no forces on us because the force of gravity toward the Earth, being unopposed by pressure on our feet or bottoms, generates solely an acceleration toward Earth, and we have that falling feeling in our stomachs as a result.

If you are in orbit, gravity makes you continuously fall towards the surface of the central body but, at the same time, you are moving sideways at such an (orbital) speed that the surface of that body curves away from you at just the same rate as you fall toward it. So, you never hit the surface (eg Earth) and never experience any gravity or feel your weight because you are in free fall under the effect of omnipresent gravity.

So, why is space a near vacuum?

Astronomy is sometimes defined as the scientific study of all things beyond the Earth's atmosphere, beyond being in the sense that effects of the atmosphere (like drag on orbiting satellites) are negligible. In reality, this happens pretty close to the Earth, essentially because the air is quite cool and the thermal motion of most air molecules is too slow for them to rise very high before gravity pulls them back. If this were not true, the atmosphere would quite quickly leak away into space. In fact, the density of our atmosphere falls by a factor of about 2.7 for every 10 km we rise. Ten km is roughly the height of Everest or the flight altitude of some aircraft, hence the need for oxygen when climbing or flying at this height. Applying that same factor of 2.7 each 10 km shows that, at 100 km, the air pressure is already around 20,000 times smaller than at sea level. Thus, artificial satellites orbiting well above this altitude avoid the air-drag present at lower levels. Therefore, the near vacuum of space is not because there is *no* gravity but because there *is* gravity which confines atmospheric gases very close to planets and stars.

In interplanetary space near Earth, the density of matter quickly becomes extremely small, mainly not from our outer atmosphere but from the hydrogen gas of the thin solar wind (Section 3.1) that blows out from the Sun – which is made up of about one atom per cm^3 (about the volume of a sugar cube). This is more than 10^{21} times less than air at Earth sea level and 1,000 times less than can be achieved in the very best man-made laboratory vacuums! The density in intergalactic space and the average density of the whole Universe is much tinier still. These are near perfect vacuums, almost devoid of ordinary matter, though modern physics involves the curious concept of vacuum energy comprising so-called virtual particles. These can materialise under extreme physical conditions like in the intense gravity of a black hole or by collisions of very high energy particles in particle accelerators like that at the European Centre for Nuclear Research, Geneva (CERN). It is in such places that physicists hunt for the exotic particles which underlie the nature of matter and the forces within it. One such is the Higgs Boson particle, which was predicted as the source of mass in matter in 1964 by Peter Higgs of the University of Edinburgh. Its first detection at CERN, announced on 4 July 2012, won Higgs a share in the 2013 Nobel Prize in Physics.

However, returning from that weird sub-nuclear world of bosons and quarks (a name drawn from Joyce's *Finnegan's Wake*), we can ask what a vacuum is in everyday terms. Strictly speaking, it is a region where the density of ordinary matter is zero. In practice, it means that the density is so small as to have negligible effects on anything in which we are interested. Among the everyday physical effects present in matter which are negligible in a near vacuum are gas pressure and some forms of heat transport.

For those unfamiliar with the science jargon, heat is the energy contained in the random motion of particles in materials. In the case of a gas, this is the kinetic energy of its randomly moving free particles. Heat energy per particle defines the temperature (T), while heat content per unit volume and heat flow also change proportionately to how spatially dense the material is (particles per unit volume). In physics, the scale usually used for temperature is the kelvin (K) scale, named after Irish-born William Thomson (1824–1907), later Lord Kelvin, Glasgow Chair of Natural Philosophy for over 50 years. A temperature interval of 1 kelvin (1 K) is equal to an interval of 1 celsius (1°C), but has a much lower zero point, such that 0 K = -273.15° C (absolute zero), the temperature at which random thermal motion ceases entirely. For very hot objects like the Sun, the values of the temperature in K and in ° C are roughly the same. For instance, the Sun's surface T value is about 5,800 K or 6,073° C, while on the Fahrenheit (° F) scale (rarely used in science, except in the US), its value is about 1.8 times higher.

Heat is spread around our homes from our so-called radiators but mostly by convection (and some conduction) involving flow of air molecules from hot to cold places rather than by (infra-red wavelength) electromagnetic radiation. If you take the air away to leave a vacuum, the only way left for heat to travel is by electromagnetic radiation, which is able to propagate across the vacuum of space, a fact predicted by James Clerk Maxwell (Section 5.8). This was an important fact in the invention by the Scot James Dewar in 1892 of the Dewar vacuum (aka Thermos) flask used to keep liquids hot or cold. These comprise a glass bottle with double walls and a thin vacuum gap to prevent heat loss to the outside world via conduction or convection, and additionally with a highly reflective coating to prevent the escape of heat by radiation. On the other hand, in the vacuum of space, *only* radiation can transport energy and such transport is what enables our luminous Sun to keep us warm 150 million km away on Earth.

Returning to the subject of space and orbits, one of the most exciting developments during the writing of this book has been the rapid and successful progress of spaceflight programmes led by private commercial bodies, most notably Elon Musk's SpaceX. Musk's ambitious programme is partly aimed at getting people into orbit and ultimately onto Mars using his modular reusable heavy launch rocket system. His spectacular recent successes in the controlled return of the first and second stages of his reusable launch system to land upright on their launchpads paled into insignificance compared with the successful 6 February 2018 launch of his deep space heavy launch Falcon rocket. The final stage of this rocket launch system, the most powerful currently available in the world, is now speeding its way along an orbit that will take it beyond Mars, carrying Musk's own midnight cherry red Tesla Roadster electric super car as whimsical payload, with a dummy driver called Starman (Figure 1.2.2). Via NASA, JPL and various phone app fans are able to follow the current location and speed of Starman and his Tesla in his space trip.

Figure 1.2.2 Elon Musk's Tesla Roadster and Starman aboard SpaceX probe
(commons.wikimedia.org/wiki/File:Elon_Musk%27s_Tesla_
Roadster_(40110304192).jpg)

'Astronauts' – the nem
Leeterally means 'sky sailors' –
Watch no get press-ganged.

Before leaving the theme of Earth's atmosphere and the near vacuum of space beyond it, we should emphasise that the definition of astronomy as 'the study of things beyond it' is rather arbitrary. This is increasingly true with today's advances in exploring and observing the atmospheres of other planets and even of exoplanets. These are revealing Earth-like 'weather' and geophysical phenomena on them, such as lightning on Venus and Jupiter and possibly on Exoplanet HAT-P-11b; aurorae on Jupiter and Saturn, and even Mars (Section 3.1); plus volcanoes and geysers on Jupiter's moons, Io and Europa (Section 2.4). In fact, the Earth's interior surface and atmosphere are rich in information about science relevant today to the cosmos including the blossoming field of astrobiology, as well as to any reflection on cosmic beauty. Terrestrial water clouds alone are wonders to behold and study (see, for example, www.cloudappreciationsociety. org), as are the numerous other rarer but equally beautiful phenomena (see, for example,

Figure 1.2.3 (left) Noctilucent clouds over Glasgow; (right) over David Stirling statue, Bridge of Allan
(JC Brown, Glasgow; Douglas Cooper, Doune)

'Light and Colour in the Outdoors', M Minnaert, 2008).
Among these are noctilucent clouds (www.wikipedia.org/
wiki/Noctilucent_cloud) caused by very high altitude ice
crystals scattering sunlight well after sunset at ground level
(Figure 1.2.3) and multiple rainbows, which are occasionally seen over very calm water (Figure 1.2.4). Not only
does the water reflect the normal inner and outer rainbows
caused by direct sunlight, which have arcs shorter than semicircles, but the sun reflected in the water acts as a second
light source, creating inner and outer rainbows, which are
fainter but larger (arcs longer than semicircles) than those
made by direct sunlight.

Figure 1.2.4 Multiple rainbows,
Isle of Skye
(MI Brown, Glasgow)

Musky's Motor Caur

Tae gie his Falcon rocket 'Mass',
An pit it tae the test,
Bold Elon Musk said, 'Dinnae fash!
Ah've juist the thing!' – ye've guessed!
As doun the freeway syne it flashed,
Fowk gecked an stared, impressed!
'Ah best mak shair ah dinnae crash! –
Get oot the road ya pest!'
    Toot! Toot!
  Gaed Musky's motor caur!

Till oan the launch pad there it sat,
Secured bi bungy straps!
The gluve boax suin he redd o tat,
An in't he stowed a map!
'Aince oot o range o Intelsats,
We'll no can uise wir Apps,
Whiles syne the batteries wull gae flat,
When temperatures hae drapt;
    Nae probs!'
  Fir Musky's motor caur!

The Starman in the drivin sate,
He gien a cheeky wink,
Syne raised his thoomb, (he wisnae blate!)
His spacesuit luiked fair dink!
Tho thrusters made his arse gyrate,
He nevir e'en blinkt!
An listent tae the countdown prate,
As he sat oan the brink,
    o history!
  In Musky's motor caur!

Syne ower Cape Canaveral,
Did Elon's rocket soar,
Wi's fingers crossed, nae fireball!
It's luikin guid sae faur!
As syne the boosters fell awa,
They let oot sic a splore,
When gravity wis overhauled,
An intae space it roared!
  Tae cheers!
 Did Musky's motor caur!

He's set free nou tae gie it ping!
An flairs her tae the metal!
Whiles tae the steerin wheel he clings,
As fearlessly he ettles,
Tae mak his Cherry Red roadster sing,
Starman's in famous fettle!
An evri 'land speed' record's dinged –
Wi ne'er a drap o petrol!
  It's true!
 In Musky's motor caur!

Oot past the mune, her siller licht,
Reflecks the Tesla's chrome,
The Stars an Stripes are gien a flicht,
Whaur Airmstrang aince cried home,
Nae need tae gie windscreens a dicht,
As roond the starns ye roam,
Mind thon rid planet in yer sichts,
An keep oan keepin oan...
  An oan!
 In Musky's motor caur!

As 'Astronomical Units' flash,
He warstles Time an Space,
Starman tae his mast is lashed,
An primed tae win the race!
Inscrutable, eternal, gash,
He stares intae the face,
O Heivin, fully unabashed,
Fir syne he wull embrace,
  A secret,
 In Musky's motor caur

Sae whaur's he nou oan his cosmic tour,
Three million miles he's clocked,
At seven thousand miles an hour,
Twelve pynts, an licence dockt!
Ageless Starman, coorse set shure,
Haundbrakes round cosmic rocks,
His helmet's visage, grim an dour,
Reflects as he taks stock,
  O Mars!
In Musky's motor caur

1.3 In the Beginning

Figure 1.3.1 *Big Bang Flash*
(Lynette Cook, San Francisco extrasolar.spaceart.org)

It seems like a good plan in the beginning of a book about oor big braw cosmos to include a few thoughts about the beginnings of the cosmos itself. According to current mainstream science, the whole of our cosmos originated 13.8 billion years ago as an entity of almost zero size, but of immense heat and density expanding at near light speed, in an event called the Big Bang. Figure 1.3.1 shows a portrayal of the Big Bang's early phase in a small extract of great space artist Lynette Cook's superb portrayal of the whole Big Bang Universe. *The Story of the Universe* was created for her parents and contained all stages of cosmic evolution to life itself. The segment here, a masterpiece in itself, shows the early Universe only out to when nuclei and atoms formed.

As it expanded, the mean cosmic density (mass per unit volume – kg per m³) and temperature (mean particle energy) fell, extremely quickly at first (on timescales like 10^{-40} sec), slowing steadily to its presently very slow rate (timescale over 10^{17} sec) in the present Universe. The nature of the matter comprising the cosmos, and the laws of physics relevant to its evolution, changed enormously as time went by. This is basically because the four distinct physical forces via which matter interacts today – gravity, electromagnetism, weak and strong nuclear forces – are thought to have been squashed into one in the earliest stages of the Universe, and became disentangled as time went by. We summarise this development below, after we clarify a couple of general points about space and time in the Big Bang scenario. These are so far from our everyday experiences that they are very hard to grasp other than in mathematical language, though analogies can help.

Firstly, the tendency to think of the Big Bang expansion of the cosmos as resembling a 'normal' (bomb-like) explosion is misleading. A bomb explodes at a specific time in a definite place in a space and expands *into* something – be it the air, or water, or empty space. This is not the case for the Big Bang, prior to which nothing existed, including matter, space and time itself. In the Big Bang, 3D space came into being rather than expanding into a pre-existing 3D environment around it. An analogy for this is the 2D surface of an everyday balloon stretching as its volume expands into the third dimension of ordinary 3D space. This is not a very easy concept to grasp. Even harder is the concept of time itself beginning at the Big Bang – hard because, in everyday experience, anything that begins does so at a specific time but what can that mean if time does not yet exist? Words like 'yet' and 'exist' as yet have no meaning. How can the cosmos know how to expand at time zero given that physical laws cannot already exist to govern it, as there is no 'already'. Anyhow, once space and time *do* exist and space expands with the passage of

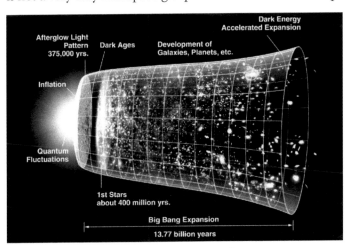

Figure 1.3.2 Schematic Timeline of Big Bang
(NASA WMAP Team commons.wikimedia.org/wiki/File:CMB_Timeline300_no_WMAP.jpg)

time, the beginnings of our braw cosmos can be divided into several very distinct phases, as hugely diverse in their duration as they are in their physical nature.

The Very Early Universe (Planck and Inflation Eras)

This interval of time, from 10^{-44}–10^{-33} sec, is absurdly small by any human standards, but was packed with high speed action. The cosmic density, temperature and particle energies were vastly higher than anything known in the laboratory or even in particle accelerators like CERN. So, physics theories of that era are, at best, highly speculative. In particular, in the very earliest stages (10^{-44} or 10 billionths of a billionth of a billionth of a billionth of a sec) post-Big Bang (PBB), all of nature's four distinct forces (gravity, electromagnetism, weak and strong nuclear) are, as already mentioned, thought to have been squashed into one force and dominated by quantum effects. No satisfactory theory yet exists of quantum gravity to allow adequate description of this state. Secondly, in order to explain various features observed in today's cosmos, in particular the smoothness of its structure on large scales, it has been conjectured that during the period up to about 10^{-33} sec PBB, the size of the cosmos must have undergone a vast burst of very fast expansion (10^{60} times) – so-called cosmic inflation – driven by some huge positive energy field conjectured to exist. We have no other laboratory or observational evidence for such inflation but it seems to fit some key facts, as did epicycles in mediaeval models of planetary motion and as does the dark energy field dreamt up as an ad hoc (*Deus ex Machina*) 'explanation' of the recent discovery that expansion of the cosmos accelerates in its old age (Chapter 8).

The Early Universe (Particle Creation/Radiation Epoch)

As the expansion proceeded, by around 1 sec PBB, there was progressive decoupling of the force types and appearance of fundamental particle types, including quarks, electrons, photons, neutrinos and more exotic types than protons and neutrons. By 10 sec PBB, the photon epoch started, in which electromagnetic radiation dominated the energy of the still very hot Universe with some nucleosynthesis occurring in its first few minutes. For the remainder (~ 400,000 years) of this epoch, the Universe contained a hot dense plasma of charged particles (nuclei, electrons) and light (photons), but too hot for neutral atoms to form by recombination of nuclei with electrons. This hot, but rapidly cooling, plasma scattered and strongly absorbed all radiation so the Universe was opaque.

Recombination Era/ Dark Ages

At around 300,000 years PBB, the temperature had become low enough (~ 3,000 K) for electrons and nuclei to start recombining into neutral atoms, allowing the large-scale cosmos to become mostly transparent. The resulting cosmos-wide burst of recombination radiation at visible wavelengths was ultimately stretched by cosmic expansion to become what we now see as the cosmic microwave background radiation (CMBR) remnant of the Big Bang. (You can hear CMBR as part of the hiss from a TV when it is tuned off-channel.) It also cooled the atoms which settled into low energy levels that emit very little radiation, the cosmos sinking into what has been termed its Dark Ages, lasting some 100 million years. During this time, matter continued to cool not only via expansion but also via loss of heat by radiation through its transparent surroundings and the cosmic gas pressure fell.

Structure Formation (Galaxies, Stars and Planets)

Until around 100 million years PBB, the temperature was high enough that gas pressure prevented density clumps from shrinking under their own gravity. Beyond that time, cooling allowed formation, by such shrinkage, of structures on all scales, from clusters of galaxies to galaxies, star clusters, stars and planetary systems. See Sections 2.1, 3.4, and 4.2 for more on these but noteworthy milestone commencement dates – measured Before the Present (BP), not PBB – include: Milky Way disk ~ 5 billion years BP; Sun and Solar System ~ 4.6 billion years BP; and the first traces of life on Earth ~ 3.5 billion years BP.

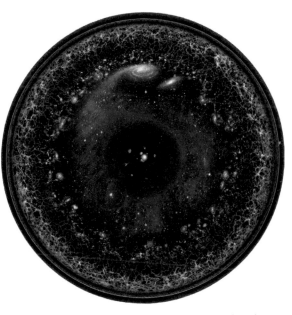

Figure 1.3.3 Observable Universe Logarithmic digital art
(Carlos Budassi 2012. www.facebook.com/pablocarlosbudassi/)

Reviewing the above, one sees that key stages in cosmic evolution occurred in time steps and space scales which became larger and larger, not linearly like 10, 20, 30… but geometrically (or exponentially) steps like 10^1, 10^2, 10^3… To display such steps, taking us all the way from 10^{-44} sec to 10^{10} years (3×10^{17} sec) and, correspondingly huge size steps, we use a *logarithmic scale* on the display such that each factor of 10 corresponds to constant steps. A lovely insightful artistic example of this is *The Observable Universe Logarithmic* by Pablo Carlos Budassi shown in Figure 1.3.3.

CHAPTER 2

Planets and Planetary Systems

Whumlin roond the Sun,
Whumlin aroond evri Sun,
Whit's it aa aboot?

2.1 What are Planetary Systems?

BASICALLY, A PLANETARY system comprises a rather small suite of planets together with a large quantity of small debris of various types, all of which are lighter and cooler than the hot self-luminous star (Chapter 3) they are orbiting – in our case, the Sun (Section 3.1). The smaller bodies range in mass down from dwarf planets (eg Pluto) and minor planets (asteroids, eg Ceres) to comets and small rocky/icy/dusty particles and are mainly seen by means of reflected starlight.

Our own planetary system, or the Solar System, is centred on our star (the Sun) and is

Figure 2.1.1 (left) Solar System Montage of eight planets, Sun behind, and minor bodies in front
(NASA JPL photojournal.jpl.nasa.gov)

Figure 2.1.2 (right) Planet Suite inspired by a BBC broadcast of Holst's Planets Suite, 20/09/07
(JC Brown, Glasgow)

Figure 2.1.3 Saturn afloat in a giant water bath
(Fotolibra. Image © Miles Kelly)

comprised of eight planets, namely, in order from the Sun: the four small inner rocky terrestrial planets – Mercury, Venus, Earth (and its relatively huge Moon) and Mars; and the four outer (~ 10x) larger Gas Giants – Jupiter, Saturn, Uranus and Neptune. Jupiter is the largest and outweighs all the other planets added together, while Saturn is the least dense. In fact, it is less dense than water and would float in water if one had a huge enough ocean or bath to put it in (Figure 2.1.3).

The small terrestrial planets are hot enough and have low enough gravity for the fast-moving gas molecules of light elements, especially hydrogen (H) and helium (He), to escape but cool enough for heavy elements to be solid or liquid. Thus, these planets are mainly small dense solid rocky bodies (some with a molten core) surrounded by a geometrically thin gaseous atmosphere (plus a liquid ocean in the case of Earth). Two of them – Earth and Mars – are orbited by rocky moons.

The four outer Gas Giants are massive and cool enough to have retained light elements, and are mainly made up of a large gaseous body of H and He that surround a compressed liquid/solid core. The Gas Giants are all orbited by rings of small particles (Section 2.5) and by numerous moons of diverse sizes and properties (Section 2.4). The cold solid dwarf planets mostly orbit well outside Neptune, while the smaller solid asteroids are more widespread, but with some concentration in belts such as between the orbits of Mars and Jupiter. Large (up to 10 km or so) icy, dusty boulders (comets) most of the time move slowly at huge, near interstellar distances from the Sun. However, they are in very long highly eccentric (non-circular) orbits and periodically fall inward and approach the Sun, heating and vaporising to form awe-inspiring huge tails made of dust, water vapour and other volatiles. Loose dusty debris, including that from comets, contributes to random shooting stars (meteors) and to the diffuse zodiacal light.

PLANETS

Prior to the discoveries of Uranus (1781) and of Neptune (1846), only five planets were known beyond the Earth and, until the mid-1900s, our knowledge of all the planets was quite limited by the distorting effect of the atmosphere on images of them, technically called atmospheric *seeing*. This is due to the refraction (bending) of light in rising cells of warm air. It is what causes stars to twinkle and distorts images of the planets and other bodies. In the last half century or so, this situation has been radically transformed by advances in ground-based telescope technology (adaptive optics) and computational image analysis. An even bigger impact has been from the advent of Earth-orbiting telescopes like the Hubble Space Telescope (HST) and of interplanetary space probes conducting close-up imaging and measurements and, in some cases, delivering surface landers. In that half-century, probes have visited every planet in the Solar System (even the dwarf planet Pluto), while in the last quarter-century, telescopes (mainly space-borne) have discovered thousands of exoplanets around stars other than the Sun (Section 2.6).

Jupiter Images

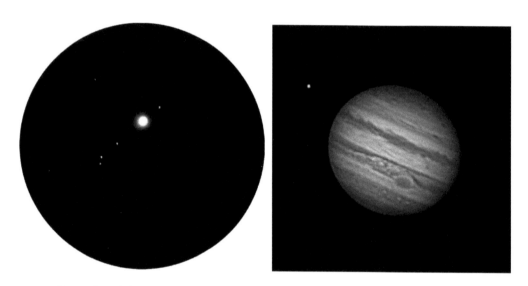

Figures 2.1.4 (left) and 2.1.5 (right): Jupiter and some Galilean moons in modern binoculars
and imaged with advanced amateur facilities
(Steven Gray www.cosmosplanetarium.co.uk)

Figures 2.1.4–2.1.7 illustrate the huge differences between images of Jupiter, first seen through mid-range modern binoculars – already far better than the first views by Galileo through his tiny telescope – then as taken by modern amateur astrophographers, and on to those shown by HST and interplanetary probes like *Voyager*, *Galileo* and *Juno*. Advances in imaging technology and processing have enabled today's amateur imaging of the planets to become superior to that achieved between 1950 and

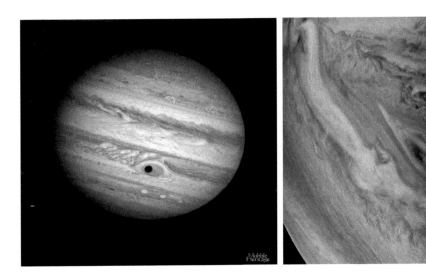

Figure 2.1.6 (left) Jupiter and Ganymede shadow as seen by HST
(NASA/ESA/HST Hubble Heritage)

Figure 2.1.7 (right) Jupiter jetstream from the *Juno* space probe
(NASA/JPL/ JUNO -Caltech/SwRI/MSSS/Gerald Eichstad/Sean Doran)

1970 in the heyday of the Palomar Observatory's 200-inch telescope (the world's largest at that time). For example, the superb image of Jupiter with its Moon Europa (Figure 2.1.5) by Steven Grey of Cosmosplanetarium was based on recording 10,000 frames over a period of around three minutes using his Sky-Watcher Skyliner 300P telescope. The best 3,000 of these frames were selected, then stacked and blended into a single image using specialised software. Stacking is the superposition of a sequence of these good images, reducing random fluctuations by making them into an average version. Blending is the result of image processing algorithms which further enhance image quality. This process minimises atmospheric seeing and other blurring effects, rendering visible far more detail than can be seen in one single frame. Today's large ground-based telescopes utilise such stacking techniques but enhance images further by real-time use of so-called adaptive optics, where the shape of a multipiece mirror is rapidly adjusted by moving its pieces to cancel out atmospheric seeing effects sampled by laser. This allows them to rival and even excel HST images at optical wavelengths.

Figures 2.1.8–2.1.11 show parallel progress in imaging Saturn, Figure 2.1.8 showing just how primitive Galileo's view was.

Saturn Images

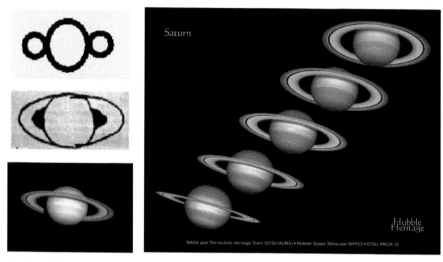

Figures 2.1.8 (top left) Galileo's sketches c.1610 : rings seen as 'moons', four years later as 'arms', 'vanishing' in between
(P. Mahaffy; NASA GSFC attic.gsfc.nasa.gov/huygensgcms/Shistory.htm)

Figure 2.1.9 (bottom left) imaged and processed with modern amateur equipment
(P. Lawrence, Selsey digitalsky.org.uk)

Figure 2.1.10 (right) at varying angles from HST
(NASA/ESA Hubble Heritage Team www.spacetelescope.org/images/opo0115a/)

Figure 2.1.11 Saturn from the NASA/ESA *Cassini-Huygens* space probe
(ESA/NASA/JPL photojournal.jpl.nasa.gov/mission/Juno)

Mars Images

Mars, though much smaller than Jupiter, has an orbit not much larger than Earth's (1 AU). So, when Mars is closest to Earth (opposite the Sun in the sky), it has a maximum angular size that is roughly half of Jupiter's, while at its farthest it is only about one tenth of Jupiter. That variation in Mars' apparent size is enhanced by the small eccentricity (non-circularity) of Mars' orbit. Mars photographers, especially amateurs, therefore seize on favourable Mars oppositions to get the best possible images, like that by Steven Gray shown in Figure 2.1.12 – obtained using image stacking and processing like Figures 2.1.5 and 2.1.9. Getting the best image is also sometimes hampered by the occurrence of major dust storms on Mars which can hide surface features (Figure 2.1.13) and even block sunlight from recharging surface rover batteries such as those of NASA's *Curiosity*.

Curiosity landed on Mars in August 2012 and, six years on, it has vastly exceeded its two-year target mission duration, traveling thus far over 20 km on Mars. One factor in this has been the effect of high Martian winds that have helped to remove storm dust from the solar panels and have allowed it to recharge. Figures 2.1.14 and 2.1.15 (overleaf) show the *Curiosity*'s route, climbing up Vera Rubin Ridge toward Mount Sharp, near Mars Glenelg. The ridge was named as a tribute to American astronomer, Vera Rubin (1928–2016), whose galaxy research led to the concept of dark matter (Section 4.2).

Surface rovers like *Curiosity* do important in-situ measurements (of local rock structure and chemical make-up, atmosphere composition and winds etc) and close-up imaging, but

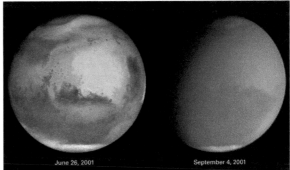

Figure 2.1.12 (left) Amateur image of Mars
(Steven Grey)

Figure 2.1.13 (right) Mars dust storm 2001
(NASA/ESA/HST)

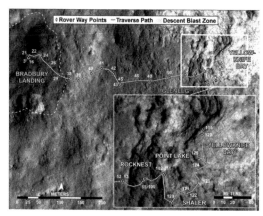

Figure 2.1.14 (left) *Curiosity* route map
(NASA/JPL-Caltech/MSSS)

Figure 2.1.15 *Curiosity* selfie on Vera Rubin Ridge
(NASA/JPL-Caltech/MSSS)

they move slowly and have a very restricted area of coverage. So, in terms of images, the most amazing and diverse images of Mars have been from the high-resolution cameras on orbiters like Mars Reconnaissance Orbiter (MRO), examples of which are shown in Figures 2.1.17–2.1.19.

Many features on Mars are named after parts of the Yellowknife area of northern Canada. Furthermore, the original Glenelg is a secluded village in the Kintail/Lochalsh/Skye vicinity of Scotland (Figure 2.1.16). The village was home to a day-long Glenelg-Mars twinning event in October 2012, which doubled the population of the village for the day. This included planetarium shows, a talk by Shuttle astronaut Bonnie Dunbar, a teleconference Q&A session with a NASA *Curiosity* mission team leader, and culminated in a ceilidh. The fact that *Curiosity* followed a to-and-fro path was allegedly linked to the feature-naming because Glenelg is a palindrome (gle-n-elg).

Nature and Rotation of Planets

As to the physical nature of planets etc, we will see in Section 3.4 that stars are fairly simple to understand theoretically because they are very hot and their material is mostly in the gaseous state. This means they have a simple form of their so-called 'Equation of State', which inter-relates their temperature, pressure and density. The same applies to at least the outer layers of the four Gas Giant planets, which are

Figure 2.1.16 Five Sisters of Kintail near Glenelg, Scotland
(JC Brown)

mainly comprised of gaseous hydrogen (though with liquid/solid cores). By contrast, the four inner terrestrial (rocky) planets and all of the moons (as well as dwarf

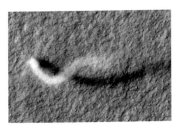  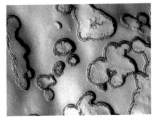

Figures 2.1.17, 2.1.18 and 2.1.19 Mars (left to right): Serpent dust devil; Hanging sand dunes; Dry-ice polar cap summer thaw
(NASA JPL MRO)

planets like Pluto and asteroids) are mostly in the solid state of matter (plus a little liquid and a gaseous atmosphere). Compared with gases, these have much more complex Equations of State and diverse forms of matter (for example, soot, graphite and diamond are all forms of solid carbon). This fact has given our planetary space probes lots of surprises for the rocky planets, for the moons of the Gas Giants (Section 2.5), and for comets. Comet structure involves *all* states of matter, from dense solids/dusty ices and liquids, through warm molecular gases, to hot ionised magnetised dusty plasmas, the last being hot enough to have lost some of their atomic electrons, so endowing with electrical and magnetic properties much more complex than possessed by neutral gases. While this diversity of states makes the components of planetary systems much harder to understand than gaseous stars, it is also what gives them (not least our planet) their magnificent diversity and beauty.

The Gas Giants, though varied in size, are fairly similar to each other, all having short rotation times of less than one Earth day, which centrifugally flattens them to oblate (tangerine) shapes. This includes Uranus, but its rotation geometry is very strange. It orbits the Sun in about 84 Earth years and spins once per 18 hours or so, but the axis of that spin is almost in the plane of its orbit rather than near perpendicular to it, as is the case for all other planets. The axial spin period is usually what determines a planet's day and night length but, in the case of Uranus, the Sun remains above the (rotational) equator hemisphere for half of the orbital period – ie 42 Earth years – then remains in the opposite hemisphere for the other half orbit. During its long orbital night, the side of Uranus away from the Sun becomes the coldest place in the Solar System.

The other strange planetary rotation is that of Venus, which rotates relative to the stars only once per 243 days but it only takes 225 days to orbit the Sun. So some might say, and often do, that:

On Venus wan day
Lasts oot fir mair as a year;
Get yer heid roond that!

However, it's not quite that simple because Venus's 243 (Earth) day spin is the 'sidereal day', ie the time between features aligning with the same stars. The time to align with the Sun again (the solar day) is different, because Venus orbits the Sun. The other complication is that Venus is spinning clockwise (as seen from the north), not anticlockwise like all the other planets. The upshot of all this is that the Venusian solar day is 116.75 Earth days or roughly half a Venusian year.

Over the last two decades the number of planets known has increased several hundred-fold due to discoveries of exoplanets around other stars. We talk about them in detail in Section 2.6 but first say here a bit about how planets and the various minor bodies that make up planetary systems form. Then in Sections 2.2–2.5, we take a closer look at our Moon, other moons, and Saturn's rings. Physically all 'planets' are bodies massive enough to shrink and be held together by their own gravity, like stars, but not massive enough to create nuclear fusion temperatures in their cores as stars do. Note that planets and dwarf planets are heavy enough for gravity to defeat rock strength and make them near spherical, while the typically knobbly shapes of minor planets and of comet nuclei are affected by material strength.

How *do* planets form?

In Section 3.4, we describe how stars form and how they became hot by the shrinking of giant cool gas clouds under self-gravity. For the formation of planetary systems, however, another very important feature of gas clouds is that they generally rotate, although very slowly. As they shrink, they spin faster, like a spinning ice-skater pulling in the mass of extended arms and legs. This conservation of angular momentum means that the rotation time is proportional to the square of the body's effective size. This is called the radius of gyration. A spinning skater who halves their effective size by pulling in their limbs will spin four times faster. For a protostar gas cloud, the factor is enormously bigger. Even shrinking from the size of the Solar System to the size of the Sun yields around 100 million times spin-up. This would reduce a 10-year spin period to 3 seconds but for the fact that, long before then, the centrifugal effect of spinning removes matter from the outer protoSun to form a flat protoplanetary disk (Propylid).

The warm gas and dust disk orbiting the star eventually cools and forms debris of many shapes and sizes. This long-standing solar nebula concept of how the Solar System formed, and how exoplanetary systems might form, now has strong support from the direct observations of such disks around newborn stars. Propylids were first observed as small local features in young star forming gaseous nebula (Section 3.4) like Orion but are now being imaged in more detail, such as the fine example in Figure 2.1.20 of HL Tauri as seen by the giant ALMA millimetric wave telescope array, whose wavelength range penetrates the dense dust and gas disks where protoplanets form. The gaps in this disk are believed to be created by growing protoplanet formation accreting the material around their orbits.

Figure 2.1.20 ALMA protoplanetary disk of HL Tauri
(ALMA [ESO/NAOJ/NRAO], NSF)

Just how the debris cools, condenses, collides and accretes into the range of rocky, icy, liquid and gaseous forms and sizes we see in our Solar System, and today even more so in stellar exoplanetary systems, is still far from well understood. Key factors probably include migration inward and outward of proto/exoplanets as part of their history to explain, for example, the rather widespread presence of massive exoplanets located close to their stars (Hot Jupiters). In any case, the huge and growing wealth of data on exoplanets (Section 2.6) is transforming our perception and understanding of planets and of the various interesting forms of debris left around their systems, some of which we now describe.

MINOR BODIES OF THE SOLAR SYSTEM

Asteroids

Asteroids (also known as minor planets) are, like the terrestrial planets (Mercury, Venus, Earth and Mars) and the smaller dwarf planets (like Pluto), mainly solid/rocky bodies, although some show evidence of the presence of some ice or water. However, asteroids are smaller, much more numerous and spread across substantial areas of the Solar System, and more irregular in shape because their masses are too small for gravity to crush them into near spherical form. The weakness of their gravity also makes them unable to retain any significant atmosphere.

The first asteroid discovered was Ceres in 1801. At around 1,000 km across, it is the largest asteroid in the belt between Mars and Jupiter, large enough to be quite spherical, and lies on the borderline between minor and dwarf planet status. Most asteroids are considerably smaller and, for a long time after Ceres discovery, they were mostly only recognised by their trailing motion in star tracking images. Once their orbit was deduced, their apparent brightness gave a lower limit to their size, while light variations showed them to be rotating and indicated their asphericity (deviation from a spherical shape). Progress made in telescope size and technology eventually allowed direct imaging of modest sized asteroids but today they have joined the realm of bodies visited and even landed upon by our interplanetary probes.

At the time of writing, several very exciting things were happening in the space exploration of asteroids (Figure 2.1.21). One was the approach to Asteroid 162173 Ryugu by Japan's ISIS *Hayabusa* 2 mission to deploy several surface experiments. At that time, its several small (frying pan-sized) robots were hopping on the surface of Ryugu and looking for possible landing sites among its boulders for the main probe to obtain surface samples and return them to Earth. A second event was the close encounter of NASA's *New Horizons* mission with the small body known as Ultima Thule at 1.6 billion km beyond Pluto, *New Horizons*' primary destination. This encounter resolved some of the puzzlement over the light curve and other properties of Ultima Thule. It turns out to be strongly bi-lobed in structure, like a snowman, very likely created by gentle impact of two roughly spherical objects at an early stage of Solar System formation. Gravitational attraction of the two lobes is adequate to hold them together despite the overall rotation. The shape of the rocky Ultima Thule is similar to that of the much smaller icy Comet 67P (Figure 2.1.25) although, in the latter, material forces are likely to be more important than gravity in lobe cohesion. Finally, just days after this, another stop-press piece of news came in of an asteroid suddenly developing a tail, suspectedly due to collision with another smaller asteroid.

Figure 2.1.21 Asteroids (top)
162173 Ryugu
(ISIS, JAXA Hayabusa2 Team)
Ultima Thule (bottom)
NASA *New Horizons* Mission
(NASA, Johns Hopkins APL, SWRI)

It is widely known by the public from movies like *Deep Impact* and *Armageddon* that even modestly sized space rocks can cause huge damage to the Earth's surface and to life. Impact of an object just 10–15 km across in Chixulub, Mexico around 65 million years ago is believed to have wiped out the dinosaurs, while objects 10–100 times smaller (so 10^3–10^6 times lighter) suffice to explain the Siberian explosions of 1908 (Tunguska) and 2013 (Chelyabinsk). Less well-known is that the latter, about the size of a large bus, would have wiped out the city's population had it had a near vertical trajectory through the atmosphere rather than its actually very shallow one. Such events were very much more common in the early days of the Solar System – known as the Heavy Bombardment era – when impacts sculpted the cratered surfaces of all the terrestrial planets and of our Moon, though the craters on Earth and Venus have mostly been eroded away in their dense atmospheres. Some notable exceptions are the mile-wide young (100,000-year-old) Barringer Meteor Crater in Arizona, caused by an object in the small asteroid/very large bolide meteorite range. (To be pedantic, Meteor Crater is a misnomer since meteors large enough to impact are meteorites and objects big enough to create long-lived craters are really asteroid sized.)

A huge early impact in the area of the Pacific is also thought to have led to our planet having such a large Moon. Presumably it is easier to create a large body in that way than by slowly accreting/aggregating many small ones. The reason the effects of impacts are so devastating is that the rocky impactors have very high speeds (typically 20 km/sec or 45,000 mph), at which even a 20 m (30,000 tonne) rocky object has a kinetic energy equivalent of around a megatonne of TNT. Even today the number of modest-sized asteroids is huge, but they mostly move in several orbital bands such as that between Mars and Jupiter, with only around 1,000 larger than 1 km that cross the Earth's orbit being potentially serious threats to life.

Figure 2.1.22 Barringer (meteor) crater (1.2 km across, 170 m deep) near Flagstaff, Arizona
(Julian Gibson, Glasgow, with reproduction permission of Barringer Co VP)

Modern telescopes and radars have vastly increased the number of asteroids known to us, and our knowledge of their whereabouts. The odds of Earth colliding with one of these (ie of the Earth being at the point in its orbit simultaneous with the asteroid crossing it) are a few per cent in the next 1,000 years. On a more upbeat note, however, impacts by asteroids and comets have also been suggested as possible sources of at least some of Earth's water, some organic compounds relevant to life, and even bringers of life itself according to believers, like Sir Fred Hoyle, in Panspermia theories that life has spread across the cosmos.

Comets

Like asteroids, the nuclei of comets are chunks of cold debris left over from the formation of the Solar System. However, they are generally much smaller (under 20 km or so) and much less robust, being looser and less dense aggregations of rocks, dusts, and ices formed in the very cold outermost regions of the Solar System. This hypothesised region, called the Oort Cloud, is thought to lie at a substantial fraction of the distance to the nearby stars. When the orbit of one of these dusty snowballs/icy dirtballs is disturbed, it falls unchanging toward the Sun over many thousands of years until the increasing sunlight starts to vaporise the ices. The action of sunlight and the solar wind rapidly blow this dusty vapour from a nucleus less than 10 km wide into a huge tail of water vapour, gas and dust stretching hundreds of millions of km across the Solar System (just as the steam from a kettle without its lid on fills a kitchen). For larger comets like Halley's (last apparitions in 1910 and 1986), Hale-Bopp (1995–97) and McNaught (2007), the sight of their huge tails glowing in the sunlight as they arc across our skies aroused much fear as well as awe and curiosity (Figures 2.1.23 –2.1.24). Halley's work successfully predicting the return of an earlier comet was an important vindication of Newton's theories of gravitation and orbits. In recent decades, close encounters with and landings on comets by space probes like ESA's Rosetta-Philae on the nucleus of Comet 67P/Churyumov-Gerasimenko to study their near-pristine primordial stuff has helped vastly in our understanding of planetary systems and the discovery of complex organics on them may be a further step in astrobiology.

The steep cliffs visible on 67P are over 1 km high, about the height of Ben Nevis and a tenth of Everest, and around one quarter of the comet nucleus size. (For comparison, Everest is one thousandth of the Earth's radius.) They thus look like huge, dangerous terrestrial cliffs but, with the comet's gravity only being 0.0001 of Earth's, a free-fall from their summits would actually take around 15 minutes and would impact the ground below at about 1.4 m/sec or 3 mph. On Earth, free-fall would only take around 10 sec at a

Figure 2.1.23 Comet McNaught seen from Paranal
(ESO/Sebastian Deiries, Wikimedia Commons)

Figure 2.1.24 Comet Hale-Bopp looking northwest from Powell Butte, Oregon, April 1997
The Pleiades cluster (Section 3.3) is between the Moon and Hale-Bopp
(Tequask, Wikimedia Commons)

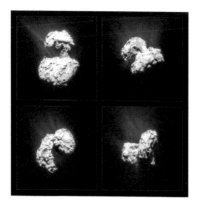 

Figure 2.1.25 (left) ESA Rosetta mission image sequence showing rotation and outgassing of Comet 67P
(ESA/Rosetta/NAVCAM, CC BY-SA IGO 3.0)

Figure 2.1.26 (right) Rosetta mission image of the steep cliffs on Comet 67P
(ESA/Rosetta/NAVCAM, CC BY-SA IGO 3.0)

speed of about 300 m/sec or 600 mph (although limited in practice to about 150 mph by air drag). So 67P is a great place for wimps like me to ski and go rock- and ice-climbing.

Meteors

Meteor is the scientific name for the shooting star phenomenon of a short-lived streak of light like a star falling from the sky. We now know these to be caused by very small particles called meteoroids (mostly sand-grain-sized). These are as fast-moving as asteroids, but most burn up in our atmosphere at an altitude of tens of kilometres, leaving a glowing streak of hot ionised atmospheric atoms which we see (and some wish on) as a shooting star. The larger sized grains/pebbles among these can be scarily bright fireballs/bolides, ending in an explosion of light as the body disintegrates into many smaller ones. Larger meteoroids (0.1–1 m depending on incidence angle) do not burn up entirely before hitting the ground as meteorites. These are rarely found, even in deserts and ice fields, and more rarely still in other terrain, the famous 1804 High Possil meteorite fall in Glasgow being one of only four ever found in Scotland. On the other hand, very small dust-size meteoroids decelerate very quickly in the air – before they can burn up – then drift slowly to the ground as micrometeorites, most easily found on the white arctic snowfields. Every night, around 10–20 random meteors are seen per hour in dark enough skies, appearing at random times and from random directions, as general parts of primordial Solar System debris. In and around the light of cities, sporadic meteors mostly go unnoticed.

More noticeable and interesting are the annual meteor showers, which very occasionally become meteor storms. Meteor showers occur when the Earth crosses the dusty residue left behind by the orbit of a comet. When that happens, the meteors all tend to come from a single direction in the sky towards which the Earth is moving, just as driving into a snow shower makes the flakes all seem to diverge out of one point. This is called the meteor shower radiant (Figures 2.1.27– 2.1.28). The specific constellation in which the radiant lies gives its name to the shower: the mid-August Perseids (which happen to peak on the Glorious Twelfth of hunting season fame) come from the constellation Perseus, the mid-November Leonids from Leo, and mid-December Geminids from Gemini.

Two very fine images taken during recent Perseid showers are shown in Figures 2.1.29 and 2.1.30. The first was a Group winner in the Astronomy Photograph of Year

Figure 2.1.27 (left) Perseid shower long exposure photo with star tracking
(NASA www.nasa.gov/topics/solarsystem/features/watchtheskies/perseid-meteor-shower-aug11-12.html)

Figure 2.1.28 (right) 'Snow Shower' demo of meteor shower radiant the car motion through the snow flakes replacing Earth motion through a cloud of meteoroids
(JC Brown)

Competition and NASA Astronomy Picture of the Day (APOD) on 13 December 2013. Note the two non-Perseid meteors at the top right and left of Figure 2.1.29 that do not come from the Perseid radiant. The images making up this composite have been manipulated to remove Earth rotation effects. The second, Figure 2.1.30, was NASA Astronomy Picture of the Day (APOD) on 17 August 2018. It shows a Perseid fireball (bolide) which, like other large fireballs, left a persistent train of glowing atmospheric atoms which drifted and evolved in upper atmosphere winds, shown in the Figure by means of a sequence of time lapsed images.

Much rarer meteor storms occur at meteor shower times when the Earth's crossing of the dusty comet orbit occurs at an especially dense clump of dust (Figure 2.1.31).

For the Leonids, such major peak years have included 1799, 1833, 1966 and 1996.

Making the most of the beauty of meteor showers requires patience, a deck chair, warm gear and ideally a large warming dram – and a night near new Moon to get the darkest skies. While waiting patiently for the fast, bright streaks (lasting a few seconds) of meteors/shooting stars, you can watch out for passing artificial satellites, the most awesome of which to see is the International Space Station (ISS) with its human cargo, some of whom have been up there for many months. Most satellites appear as points of light with near constant brightness (before fading into the Earth's shadow) and moving fairly slowly across the sky (for a few minutes). Since ISS shines by reflecting sunlight

Figure 2.1.29 Perseid shower over a Mongolian lake, August 2017
(Haitong Yu, Beijing 500px.com/haitongyu/about, apod.nasa.gov/apod/ap171213.html)

and orbits not very high above the Earth, it is only visible (and only sometimes) not too long after sunset or before sunrise, since these are the only times when it catches the sun's rays. Iridium satellites can also seen by reflecting sunlight, mainly from their big mirror-like solar panels which, due to quite fast satellite rotation, flash light towards any particular observer only rather briefly but sometimes very intensely. The times and

Figure 2.1.30 Perseid shower fireball, 17 August 2018
(Petr Horalek www.astronom.cz/horalek/, apod.nasa.gov/apod/ap180817.html)

Figure 2.1.31 Leonid storm over Niagara
Painting, November 1833, by E Weiß (died 1917) published in Bilderatlas der Sternenwelt, 1888

locations in the sky of these brief but brilliant moving flares of light are very accurately predicted for any given place and it can be a fun party trick to get the kids or revelers to stare at one part of the sky and count down the final seconds till the Iridium flare. Viewing opportunities for iss and Iridium flares can be found on many websites including Heavens-Above (www.heavens-above.com).

The Auld Professor

'A Glorious 12th indeed', he thinks,
Glimmerin at the lift abune the Bens.
Nicht's daurk mantle faws upon the Cuillins,
Syne black as velvet pickt wi' gliskin pearls;
Natuir's stairk Graund Orrery,
Ne'er failin in its pouer tae dumfouner!

Aa is still, an ceptin fir the heivins naethin steers,
Till aa a suddent a timorsome deer
Nervously stegs athort the field;
Lugs twitch, she yerks her heid,
Daurk een drink in the tentless starns,
Syne swith, turns tail an dairts awa.
Swipperly she's swallaed bi the forest.

Prof patientfu leans oan the drystane dyke,
That's claithed in rime o hoary moss,
Nursin the gless o whisky in his haund,
A stiff dram o his favourite Talisker,
Bides cannie the musica universalis.

Then thair they are! Siccar as the mornin Sun!
Flashin, gliskin, glentin sparks o fire!
The Perseids, wingin their endless wey,
Athort the braid black welkin o the nicht.

He hauds the gless o whisky tae his ee,
An' keeks fir fun oot throu its pepper smeek,
The Amber o his nectar's fremt meniscus,
A skime o licht glints past his eident gaze…
'Claucht ye!'

He clamps a haund doun oan the gless,
Luiks round tae speir that nane hae seen this daftness,
Then steikin baith his een, taks aff his dram;
An' drinks the perfect fire o the cosmos.

International Space Station

'We come spinning out of nothingness, scattering stars like dust.'
– Jalaluddin Rumi

The mune, a conjuror's penny,
Kythes itsel frae ahint the Knipes,
Its muckle gowden disc
Yawns an raxes til the lift.

Fordoverit starns aa streetch an wauken,
Blinkin frae their daylicht dwams,
Tae tak their waukrife place empyrean.

Athort this heivenly canvas jee's
A bricht unwaverin steady licht,
Poured frae the globe's meniscal rim,
Traivellin seeventeen thousan miles an hour...
Haudin fast its coorse frae west til east...

A tin-can that inhauds the best o us!
American, Russian, Japanese;
Sindrie ither nations o the warld.
Think oan; a Trident system,
That handsels life insteid o dolin daith.

Inside they hae the graith,
Tae spier anent the lear o unkent things;
Growe mair crops tae feed oor stairvin planet,
Gene mutation's modren sorcery,
Engineerin eldritch cures
Tae brak the spell o deidly disease...

The eident crew, thrang wi whigmaleeries,
Tentie haunds that tend some fykie task –
Cannily; they're cairtin aa oor futures.

Amazed, ah watcht this human 'Star o Wonder',
Ayont the Scottish hills it fades an dwines,
Tae skim oot ower the waves o ilka ocean,
Serenely sclimb abune ilk mountain range,
Pyntin that road we'll gang, as yet untraivelled.

2.2 *Apollo*, Moonwalks and the First Space Selfie

Half a century ago, at 9.17 pm BST on 20 July 1969, the Lunar Excursion Module (LEM) of NASA's *Apollo 11* mission safely delivered the first humans (Neil Armstrong and Buzz Aldrin) to the Moon, our nearest neighbour in space. It touched down with only 40 seconds of fuel left. This was just 8 years and 2 months after JFK stated that his nation 'should commit... to... landing a man on the Moon... before this decade is out'. Eighteen months later he said that all space effort must 'be tied into getting to the Moon ahead of the Russians', and making the venture into a dangerous, high-stakes, high-price political race, as well as an awesome human endeavour. It required development of a rocket far more powerful than its predecessors, which were used to orbit manned capsules above Earth at about 1/2000th of the Moon's distance, as well as complex equipment for the landing and return. The first

Figure 2.2.1 *Apollo 11 Saturn V lift-off, July 1969*
(NASA archive)

ever flight of this *Saturn* V rocket (around 100 m tall and weighing 3,000 tonnes fully fueled) was only 20 months before the *Apollo 11* first Moon landing; the first flight with a crew was seven months prior to that landing; and the first ever manned flights out of Earth's orbit were four and two months prior. These great successes followed the one tragedy of the whole *Apollo* story. During an *Apollo 1* launch rehearsal all three crew sadly died in an intense fire accelerated by the unwise early use of an oxygen-rich atmosphere in the command module. Despite the hasty gamble to beat the Russians (who were, in reality, by then lagging far behind), only one of seven Moon-landing *Apollo* missions (11–17) failed when there was an explosion in lunar orbit aboard *Apollo 13*, which miraculously managed to return its crew home safely, thanks to the brilliant ingenuity of the crew and NASA mission teams.

The straight-line distance to the Moon is around 400,000 kilometres or 250,000 miles. This is roughly 20 times the distance round the globe from the UK to Australia, which takes around a month by the fastest ships and around a day by the fastest passenger planes. The *Apollo* flight time to lunar orbit was around three days. (For comparison the next step in interplanetary travel – to Mars – is hundreds of times further.

At 03.56.15 BST, watched by over 500 million TV viewers worldwide, Neil Armstrong (a Scottish clan known for forays away from home) made history as the first person to set foot on another world, followed 20 minutes later by Buzz Aldrin. During their two

Figure 2.2.2 First lunar selfie
(NASA archives)

and a half hour Extra Vehicular Activity (EVA), as well as 'taking giant steps and hoping their legs don't break' (to paraphrase The Police song), they set up an experiment package, collected rock samples, raised a US flag, took a call from the then President (Nixon) and took photographs.

But what *is* an Armstrong to do with his remaining time on the Moon once he's

Figure 2.2.3 Moon Rover
(NASA archives)

collected the rocks, raised the flag, done his experiments and perhaps dreamt of the small impact crater 50 km northeast eventually named after him? Pubs and clubs can't flourish with no atmosphere and there are no cattle on the Moon (they all jump over it!) for an Armstrong to rustle, as was the ancestral clan's wont. So, nothing for it but to take some holiday snaps. Among the most famous photographs taken on the *Apollo 11* mission is that taken of Aldrin by Armstrong. However, a strong feature in this scene is Armstrong himself reflected in Aldrin's spacesuit convex visor, making it the first ever space selfie, some 40 years before the word was invented.

Features in this beautiful and evocative image of a historic event have been invoked by fringe conspiracy theorists as evidence for the Moon landings being a hoax – what would be termed 'fake news' today. In this case, the alleged 'evidence' is that the geometry of the various shadows makes no sense if the only light source is the Sun (as opposed to several studio lights). The shadow patterns do look odd at first sight but one must allow for things like the distorting effects of reflection in the strongly curved visor. Other bogus arguments from these hoax theorists is that astronauts *en route* to the Moon would die in the high-energy particle radiation of the Van Allen belts. That potentially could have been true but the danger was avoided by using *Apollo* mission trajectories involving only short belt transit times and/or regions of weak radiation. Other conspiratorial arguments likewise fall under scrutiny and the fact is that faking evidence of flights to and landings on the Moon would be harder than landing on the Moon.

The Moon remained a pedestrian-only zone throughout the *Apollo 12* and *14* landings, but the arrival of three battery-powered Lunar Roving Vehicles with *Apollo 15–17* seemingly caused no exhaust pollution nor parking congestion. Also of interest here is that the *Apollo* lunar mission's colour TV cameras had a mechanical TV system like that invented in 1928 by John Logie Baird of Helensburgh, Scotland, with colour wheels sending field-sequential images to Earth where they were converted to electronic format.

The technical magnificence of the US NASA *Apollo* programme and the huge success of their unmanned Earth-orbiting observatories (like HST) and interplanetary probes (from *Pioneer* and *Voyager*s, through *Cassini-Huygens* and *Galileo*, to *New Horizons*) is unquestionable, even if originally in part driven by a political space race. It is, therefore, sad today that no human has been to the Moon since 1972, that no US vehicle has taken a man into space since the last Shuttle flight in 2011, and that, while the US Presidential ambition in 1961 was the expansive one to land the first human on the Moon, today it is to shun allies and to retreat behind border walls.

On a brighter long-term note, we close here on the story of *Apollo 14*'s so-called Moon Trees. Every *Apollo* Moon mission must have had its own tales to tell, though none with the drama of *Apollo 13*. One unusual tale that seems worthy of telling in the spirit of this book is that of *Apollo 14*. The *Apollo 14* Command Module pilot was Stuart Roosa, who had formerly been a smokejumper with the Forest Service. As a tribute to the Service, he agreed with Ed Cliff, Chief of the Service of *Apollo 14* at the time, to take around 500 assorted tree seeds to orbit the Moon with him as one of his personal 'baggage' items. The idea was partly to check if the week or so of zero gravity and spaceflight might affect the viability of the seeds, and to test this by planting them back on Earth to create trees as a tribute to the *Apollo* missions. Many were planted in Bicentennial year 1976 as part of the celebrations at various national landmarks, such as the White House, and some internationally. They mostly thrived alongside Earth-bound equivalents and became known as Moon Trees. Moon Tree 1976, planted at NASA's Kennedy Space Center on 25 June 1976 during the Center's Bicentennial exhibition still thrives there, as do other Moon Trees around the world and some of the progeny from *their* seeds, sweetly called Half Moon Trees.

Spacewalk

IM Neil Armstrong 1930–2012. First human to walk on the Moon.

The ultimate daunder…
Leeterally,
Gaun bauldly whaur nae man hus gaen afore!

We'll claim ye, Airmstrang, fir ane o us,
Nae maitter whit thon Stars an Stripes micht ettle.

That Borders reivin heritage,
Stounds stievely throu yer veins,

Forebears wha plashed throu muirland dubs,
Oan skinklin nichts,
Hooves ripplin an wimplin
Thon heivinly reflection.

They'd no see ony coin ava,
In your celestial raids;
Nae muckle siller in a puckle rocks!

Tho they kent the warth o her loosome beams;
Thair wull be munelicht aince agane!

Til the days o the mailit gluve wir tined,
An' the clan wis skail't tae aa the airts.

Invictus Maneo! Ye bide unvanquished.

Nou faur abune Gilnockie Touer,
An aa the Debatable Launds,
Ye cast yer een back hamewarts;

Ye tilt yer heid, the visor o yer helm,
Kythes a mirrored pictur o oor warld,
Syne turn asklent tae luik the ither wey,
Launds auld Johnnie cuid anely dream o!

Mune Trees

Loblolly Pine, Sycamore,
Sweetgum, Redwood, Douglas Fir;
Aa siccarly stowed mangst Stuart Roosa's kit.
A forest hained athin his haunds,
An launched oot intil space.

Throu the module's windaes they vizzy,
The ferlies o the spinnin yird;
The Amazon's muckle greenness,
East-Siberia's cranreuch Taiga,
The Congo Basin, Australia's Daintree,
An' Borneo's grushie rainforests…

Syne, wi *Apollo*'s sauf retour,
Stuart's seeds wir eidently taen;
Willin fingers delve intil the erd,
Ilk yin wis plantit,
Ilk wan taen ruit,
Raxin their wee bit heids tae drink the sun.

Nou aa these 'Mune Trees' maun be foun,
In Switzerland, America, Brazil, Japan,
Arboreal heroes ilka wan,
Winnin their wey back tae the heivins,
Tracing rings lik orbits in thair hairts.

2.3 Blue Moons, Super Moons and Other Moon Madness

Being our nearest cosmic neighbour and by far the most prominent feature in the night sky, our Moon attracts a great range of interests from hard-nosed science through simple misconceptions, superstitions and downright silliness. Professional interests in the Moon range from the composition of its rocks, as clues to the Solar System's history, to space engineering challenges of establishing bases there, and as a potential staging post for Solar System exploration. The Moon's considerable influences on Earth are also interesting: its gravitational pull and orbit around the Earth is the main driver of our tides and their many effects, and it stabilises the precession and wobble of the Earth's tilted spin axis, hence our seasons. For amateurs simply watching the monthly cycle of Moon phases – its wide altitude swings through the days, months and years, and its occasional eclipses in the Earth's shadow – are almost as fascinating as telescopic observation of its shadowed craters and mountains.

Among basic misconceptions about the Moon are that one can only see the Moon at night, not in daytime (see Figure 2.3.1) and that the phases of the Moon are caused by the Earth's shadow rather than the changing direction of the sunlight falling on it as the Moon orbits the Earth.

Another common fallacy is that one needs expensive gear to see lunar mountains and craters, but one can see hints of them along the junction of lit and unlit sides even with the naked eye and very well with basic binoculars and telescopes (see Figures 2.3.2 and 2.3.3). Readers should get out there and try this on the first clear night when the Moon is up. (But *do not* try the same thing on the Sun if you want to avoid being blinded.) Of course, alongside these are some of the best images today (eg Figure 2.3.4), which are taken from lunar orbiting spacecraft like NASA's LRO and show the lunar surface in exquisite detail.

A widespread but mistaken perception of the Moon, shared by many scientists as well as laymen, is that the Moon, especially when nearly full, looks markedly bigger (in angular size) when on the horizon than when high in the sky. This apparent effect – The Moon Illusion – has provoked lots of suggested explanations, such as increased atmospheric refraction or effects inside the eye when the Moon is fainter. However, research shows it to be entirely psychological and related in some way to how the brain judges differently the sizes of things in relation to their surroundings. In the case of the Moon, it may appear bigger

Figure 2.3.1 Moon in daylight setting behind Ben Ledi, the Trossachs, Scotland
(Keith Wilson, Callander)

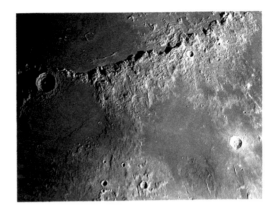

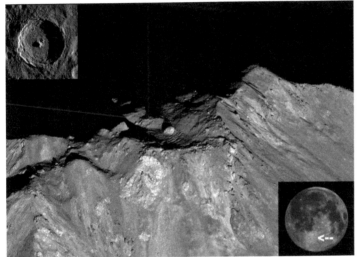

Clockwise from top left:

Figure 2.3.2
Moon craters with
phone camera at scope
eyepiece
(JC Brown)

Figure 2.3.3
Good amateur lunar
picture, image stacked
(Alex Houston, Alloa)

Figure 2.3.4
Unusual boulder atop
lunar Tycho Mountain
seen by NASA Lunar
Reconnaissance Orbiter
LRO
(NASA LRO)

when near trees and buildings at the horizon, but smaller when high up in the empty sky. A thorough investigation of this notion, proving it to be false, and exploring why we think otherwise, can be found in the book *The Mystery of the Moon Illusion* by Professor Helen Ross, formerly of the Psychology Department at the University of Stirling.

Another common mistaken notion, based on the fact that we only ever see one side of the Moon, is that the Moon does not rotate. The reality is that the Moon *does* rotate (relative to space or the distant stars), but in the same period as it orbits the Earth. So, as the Moon orbit us, it turns relative to the Universe, but keeps facing us. The physical reason is that the Earth's gravity raises a tidal bulge in the Moon and, over many millennia, the Earth's pull on that bulge has slowed the Moon's rotation so that the bulge always faces us.

There are also some lunar events, much hyped by the media and presented as if very rare, which draw much public attention, but which are in fact quite commonplace and rather minor. Most extreme of all is the allegedly rare Blue (Full) Moons, which are neither blue nor rare (despite the phrase 'once in a blue moon'). Today, a Blue Moon is defined as the second of two full Moons falling in the same month. Since the lunar phase cycle is 29.5 days long, while most months are 30 or 31 days long, one would expect two full Moons to fit into one month whenever the first one fell on the first few days of the month. This is around 10 per cent of the time and at least once a year. Blue Moons are simply a result of the arbitrariness of our month lengths being a bit longer than the lunar month and have no physical or causal significance, any more than having a birthday every four years as a result of being born by chance on 29 February.

Next in line are Super Moons which, by contrast with Blue Moons, do have physical meaning and are of some interest. A Super (Full) Moon arises when a full moon occurs while the Moon is near to its closest to Earth (perigee) in its slightly elliptical (non-circular) orbit. This makes the Moon about 5 per cent nearer than average and so looking 5 per cent larger and 10 per cent brighter in total than an average full Moon,

under otherwise similar condition. But these differences are barely detectable by eye or even by a camera – See Figure 2.3.5. The 10 per cent total brightness difference corresponds to changing a camera aperture (or f) ratio by a factor of 1.1 – eg from $f/8$ to just $f/7.62$. Furthermore, when the Moon is nearer, though its total brightness (illumination of the ground) is higher, its *surface* brightness – light per unit area of its disk – is unchanged, because its increased total light is being spread

Figure 2.3.5 Comparison of size and brightness of supermoon with average full Moon
(Shizhao, commons.wikimedia.org/wiki/File:Supermoon_size_comparison.jpg)

over a proportionately larger area of the sky. So the media hype about super moons is completely misplaced, the super moon effect in reality being much smaller than many other factors affecting the Moon's apparent brightness. In addition, super moons are not even rare, several of the 12 or 13 full moons per year being Super.

The fact is that lots of people are impressed when they see how bright the Full Moon is mainly because they rarely look at the Moon at other times. If they did, they would see that super Moons are underwhelmingly different in size and brightness from average full Moons at the same altitude and under similar atmospheric conditions. These factors which routinely cause much larger variations in the observed brightness than orbital eccentricity does. The largest of these effects is that our atmosphere absorbs and scatters a significant fraction of moonlight and starlight, and that fraction depends strongly on how near the object is to the horizon. (To give readers a feeling for angles: the angle across the width of a pinkie finger held up at arm's length is about 1°, across a fist is about 10° and across a fully open hand, pinkie to thumb, is about 25°). In the best clear skies, at sea level, this light loss is about 14 per cent overhead while lower in the sky (looking through more atmosphere), it is 26 per cent at 30° elevation and 56 per cent at 10° elevation. The mean maximum (due south) elevation of the full Moon in Central Scotland's skies varies between about 58° midwinter and 10° midsummer. So, in perfect skies, the due south full Moon here is around 36 per cent brighter in midwinter than in midsummer – an effect seven times larger than the orbital eccentricity Super Moon effect! In practice, that winter to summer variation in full Moon extinction is compounded by the summer background sky being so bright at our latitudes that the low faint full Moon is hard to notice at all.

Variations in Moon brightness over a single evening are also much larger than the Super Moon effect. On a clear winter's night, as the bright full Moon descends after midnight from 58° towards setting, it has dimmed by 55 per cent when 10° above the horizon. At middle elevations, the Moon drops around 5 per cent in brightness (the Super Moon effect) for every 3–4° it sinks in the sky. To compound all this, our atmosphere is rarely very dry or clear and contains highly variable and substantial amounts of water vapour, aerosols, dust etc, resulting in full Moon brightness changes by amounts far larger than 5 per cent from hour to hour and night to night.

So, the bottom line here is, by all means look at Super Moons but, if you just want to see a really bright full Moon, forget all the 5 per cent Super Moon hype and just choose a cloudless, clear midwinter night around midnight. To add the really small Super Moon benefit of perigee, search the internet to find out Super Moon years and dates when lunar perigee occurs around midwinter (which has the highest and brightest full Moons). The Super Moon of 1 February 2018 *was* a nice bright full one high in clear winter skies but was not noticeably different from many other non-Super full ones, such as that on 22 December 2018, seen in similar weather. It was, however, ideally placed during a night visit to the loo to give me some spectacular shots of the effects on full moonlight of rippled glass (Figure 2.3.6). The rectangular symmetry of this man-made glass reminded me somehow of the natural cosmic object called the Red Rectangle shown in Figure 2.3.7. The origin of this appearance is not well understood, though it may involve scattering from dust, and light emitted by conical gas outflows at an early stage of planetary nebula formation (Section 3.4).

Finally, having slighted everyone else for their follies, we have to confess that astronomers are also prone to their own kind of Moon-madness in their admiration of the sheer

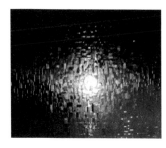  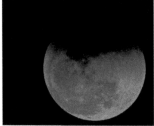

Figure 2.3.6 (left) Supermoon at 3.00 am through loo window, February 2018
(JC Brown)

Figure 2.3.7 (centre) Red Rectangle Nebula
(ESA/Hubble/NASA)

Figure 2.3.8 (right) Super Blue Blood Moon, 2018
(Irvin Calicut, Wikimedia Commons)

beauty and wonders of the sky, some of which we show in this book. Among the closest cosmic wonders in space are lunar eclipses, especially when conditions make them truly blood red (Figure 2.3.8).

Astronomers will go to crazy lengths to see these wonders for themselves, despite the challenges of weather and late hours often involved. Figure 2.3.9 shows James Green of Cosmos Planetarium braving the snow and cold in Harperrig, West Lothian to observe in the 85 per cent full moonlight of December 2017 after weeks of bad weather. A Moon-madness astro-fun time was had by all (namely James and his Cosmos partner, Steven Grey). Such a full bright moon as the one in these images has a beauty of its own but is mostly a curse for astronomers. It is too dazzling to look at directly and, being vertically lit, shows no shadows from its mountains or crater walls. All it does is to flood the sky with light which hides other cosmic sights, especially the beautiful faint diffuse nebulae (Section 3.4), the Milky Way (Section 4.1) and other galaxies (Section 4.2).

Figures 2.3.9–10 Moon-mad observers 1 and 2: Cosmosplanetarium team indulging their stargazing addiction in Baltic conditions recovering from weeks of clear sky deprivation
(Steven Gray)

Super Blue Bluid Mune

Wi its speckled ochre o a Kestrel's egg,
The mune coories in the sheddae o the yird.

Mair muckle than it's e'er bin seen afore,
Faimilies staund dumfounert in thair wunner.

Bairnies greet, owre feart tae pynt a finger,
Grandschir's luik askance at ane anither.

'It bodes nae guid' ah hear auld Minnie say,
'Cam hairst ye'll see' – nane hearken tae her glumsh.

In the midst o Callanish' michty ring,
The Shaman's antlers pynt taewart the heivins.

He casts the banes an reads whit aye wis written;
A snell wuin blaws, a bonnie lassie shivers.

2.4 The Solar System's Gobsmacking Moons

Until the discovery by Galileo (Pisa, 1564–1642) in January 1610 of Jupiter's four largest (Galilean) moons – Io, Europa, Ganymede, and Callisto – our own Moon was the only object known to orbit any planet (Figure 2.4.1). As already mentioned in Section 2.1, understanding the physics of mainly gaseous objects like hot stars and Gas Giants is easier than cold solid state or liquid bodies because their Equations of State are simpler. This is even truer for the many moons of the Solar System, as we will see.

Despite the discovery of around 180 moons since Galileo's time, ours remains the largest in the Solar System relative to its parent planet (1/4 by size, 1/81 by mass), provided one excludes Pluto (demoted in 2006 by a committee from the status of 9th planet to the first dwarf planet). Little Pluto (2/3 the size of our Moon) has no less than five moons, the largest of which, Charon, is a relative giant at over 1/2 of Pluto's size and 1/8 of its mass, making Pluto-Charon essentially a binary (dwarf) planet. In absolute size terms, Earth's Moon is 5th largest in the Solar System at around 2/3 the size of the very largest – Jupiter's Ganymede.

For unknown reasons, Earth is the only one of the inner planets to have a moon of any substantial size. Mercury and Venus have none, except ones so small as to have remained undetected. Mars has two moons – Phobos and Deimos – discovered by Asaph Hall in 1877, but they are close to the planet and only tiny. They are about 20 and 10 km across and orbit Mars in 7.7 and 30.3 hours respectively. However, presumably by a bizarre coincidence, the Laputan astronomers in Swift's *Gulliver's Travels* (1726) already referred to two small close-in Martian moons with periods of 10 and 21.5 hr.

On the other hand, the four Gas Giants all have a plethora of moons, Jupiter setting the record with 69 so far. Galileo's discovery of four moons orbiting Jupiter (like a mini Solar System) was in the same year (1610) as he observed Saturn's rings (Section 2.5). His observations of planetary and lunar features, Venus phases, and the motion of Sunspots (features mentioned in much earlier Chinese records) earned Galileo the

Figure 2.4.1 (far left) Earth-Moon system from 6.4 million km, as seen by NASA's interplanetary spacecraft *Galileo*, 16 December 1992
(NASA/JPL)

Figure 2.4.2 (left) Pluto-Charon system from 6 million km as seen by NASA's interplanetary spacecraft *New Horizons*, 8 July 2015
(JPL)

popular credit as the first person to use a telescope to discover 'imperfect' structure and changes in celestial bodies (Sunspots, lunar craters, planetary moons and rings), challenging the religious dogma of a perfect Earth-centred Universe. In fact, Thomas Harriot (Oxford, 1560–1621) had made a telescopic drawing of lunar features on 26 July 1609, over four months before Galileo.

Jupiter is huge compared with the Earth (around ten times larger) with its own complex weather systems, aurora, and giant storms like the great Red Spot (which itself is larger than the Earth). It is also 5.6 times further from the Sun than the Earth. The much feebler warming effect of sunlight alone on planets and moons at and beyond such distances means that the temperatures of their outer layers are -150° C (123 K) or less, so one might expect them all to be cold, dead places. In reality, the era of planetary exploration by space probes has revealed an astonishing variety of surface conditions and activity on Solar System moons, some offering the potential for life to thrive. Here, we briefly describe a few of the most exotic.

With around 400 active vents, Jupiter's moon, Io, is the most volcanically active of four such bodies in the Solar System (including Earth, Enceladus and Triton). Unlike Earth's volcanism, which is driven by residual internal heat and by radioactive decay in rocks, Io's derives its energy from frictional heating as the interior is tidally distorted and pumped by Jupiter's gravity as Io swings around in an eccentric orbit. Their volcanic ejecta, hurled to over 500 km high, fall back as long sulphurous lava flows, painting Io's surface red, white and black – and making it likely the stinkiest place in the Solar System. Figure 2.4.3 (left) shows active volcano plumes on Io imaged by NASA's *Galileo* probe (1995–2003). One is at top centre over Io's limb, and the other, just below mid-image near the night/day shadow line, the ring-shaped Prometheus plume rising high above Io and casting a shadow below the vent. This plume was visible in all images dating back to 1979 (*Voyagers*). Figure 2.4.3 (centre) shows active volcanic lava flow on Io imaged on 22 February 2000. Figure 2.4.3 (right) shows Io's surface becoming completely buried by these processs every few thousand years – the black and red material here being from the most recent eruptions (a few years old). This *Galileo* image shows the side of Io that is tidally locked to always face away from Jupiter.

Some of the erupting matter escapes as dust and gas into the extensive magnetosphere around Jupiter and affects Jovian aurorae (Section 3.1), whose spectra show sulphur etc.

Orbital eccentricity, and resonance with another Saturnian moon, Dione, are believed to play a similar role in frictional tidal heating of Saturn's small (500 km) moon, Enceladus, discovered in 1789 by William Herschel. In the case of Enceladus, the main product of the heating is venting through fissures in the surface ice of over 100 geysers

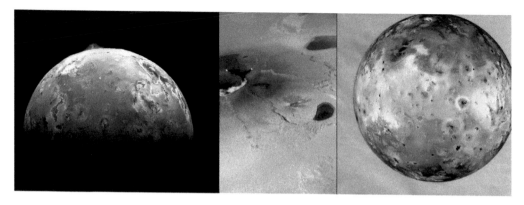

Figure 2.4.3 Io Vulcanism imaged by *Galileo* probe
Left: active plumes; Centre: active flow; Right: Io face tidally locked to point away from Jupiter is fully
covered every thousand years, black and red material here being the most recent
(NASA/JPL/SSI www.nasa.gov/multimedia/imagegallery/image_feature_764.html AND _346.html
apod.nasa.gov/apod/ap110522.html
www.nasa.gov/feature/jpl/complex-organics-bubble-up-from-ocean-world-enceladus)

discovered in 2005 (Figure 2.4.4), largely composed of water from an extensive sub-ice ocean. Some of the water vapour falls back as snow, the rest escaping and supplying most of the material making up Saturn's E-ring. The chemical composition of geyser plumes resembles that of comets, being rich in organics, so that Enceladus' hidden ocean could harbour life. Its fresh ice covering makes Enceladus one of the most reflective bodies of the Solar System making its surface far colder (-198° C) than a

Figures 2.4.4 (left) Enceladus geysers
Figure 2.4.5 (right) Europa frozen ocean surface and detail
(NASA/JPL/SSI)
www.nasa.gov/feature/jpl/complex-organics-bubble-up-from-ocean-world-enceladus
www.nasa.gov/jpl/europas-stunning-surface
www.nasa.gov/image-feature/jpl/pia20028/europa-s-varied-surface-features)

Figure 2.4.6 Callisto
(NASA/JPL/Galileo/DLR www.nasa.gov/multimedia/imagegallery/image_feature_279.html)

light-absorbing body would be there. It also shows diverse surface features, from old, heavily cratered regions to young, tectonically deformed ones. Europa (Figure 2.4.5), the smallest Galilean moon, is around ten times bigger than Enceladus but is also covered in cracked, thick ice over a liquid water ocean and, in 2016, the presence of water geysers akin to those on Enceladus was discovered. The ice shielding of such water oceans from cosmic ray bombardment makes them all even better niches for life. There are thus ambitious plans for life-seeking space-probes to visit Europa with equipment to penetrate its thick ice layer. Jupiter's largest moons, Ganymede (almost as large as Mars) and Callisto, are likewise thought to have sub-ice oceans, but Callisto's orbit involves less tidal heating, so it shows neither volcanism nor plate tectonics, and has the most impact-dominated surface in the Solar System (Figure 2.4.6).

A very different lunar beastie is Titan, Saturn's largest moon. It has long been seen

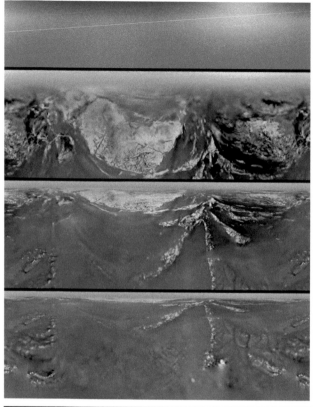

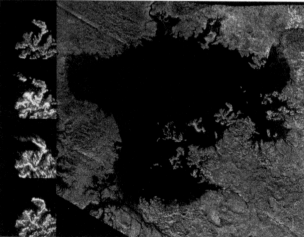

telescopically to have a significant atmosphere of low freezing point organic gases ethane (CH_3) and methane (CH_4), thought by some to be possible alternatives to water as a basis for life. This thinking was a large factor in the joint NASA/ESA funding of the *Cassini-Huygens* probe launched in October 1997 to survey Saturn and to land on Titan as the first step in its astrobiological exploration. Images of Titan taken remotely from *Cassini* in various wavelengths and locally by *Huygens* during its descent through the dense atmosphere revealed a world of many features including mountains and lakes (Figure 2.4.7–2.4.8) but of methane/ethane rather than water. A dramatised artist impression is given in Figure 2.4.9, which is one of NASA JPL's series of planetary tourism posters.

Figure 2.4.7 (top) Titan mountains and features as *Huygens* descended through the dense methane

Figure 2.4.8 (bottom) Methane lake showing feature change just right of centre

(ESA/NASA/JPL/University of Arizona/ Cassini Radar Mapper, Cornell
www.nasa.gov/content/ten-years-ago-huygens-probe-lands-on-surface-of-titan
apod.nasa.gov/apod/ap160307.html)

Figure 2.4.9 NASA/ESA/JPL Titan planetary tourism poster
(www.jpl.nasa.gov/visions-of-the-future/images/titan-small.jpg)

Figure 2.4.10 *Cassini-Huygens Slipper Orchid* JC Brown, oil on canvas board, 1998
(JC Brown, 1998)

During the *Cassini-Huygens* mission, I happened to visit an orchid fair in Glasgow and was struck by the shapes of some slipper orchid flowers as being like rounded organic versions of space probes with solar panels. I felt driven to paint the notion of a life form looking for Titan, rather than us looking for life on it (Figure 2.4.10). I realised only much later I was not the first to fuse botany and astro-technology in artistic thoughts about SETI – see Lynette Cook's piece in Figure 2.4.11, of which she says:

The huge hibiscus radio telescope is receiving a signal from an intelligent life form on a distant planet. The art represents the symbiotic relationship required between nature and technology for such a life form to survive long enough to develop a civilisation with radio technology.

Figure 2.4.11 *Cosmic Awakening*
Lynette Cook 1993 acrylic, gouache, and coloured pencil on illustration board
(Lynette Cook www.extrasolar.spaceart.org/)

The original NASA *Huygens* landing plan, like the HST optical defect, came close to being an embarrassing fiasco. This was because the tiny *Huygens* Titan-lander could only return its data to us via the *Cassini* parent probe, but the mission design failed to allow for the Doppler shift of the signal from *Huygens* as it fell toward Titan, making it undetectable by *Cassini*. However, with typical ingenuity, the mission scientists postponed and replanned the landing, so that the *Huygens* signal lay within the *Cassini* radio band throughout this first (and, so far, only) landing on a non-terrestrial moon. This mission discovered an extensive system of seas and lakes, not of water but of methane and ethane, which are liquid even at -180° C (93 K).

We close this sample of flabbergasting moon properties with one further example – Saturn's 3rd largest moon, Iapetus (Figure 2.4.12). This has a very extended equatorial ridge, making it walnut-shell shaped. Secondly, one half is 10 times darker (less reflective) than the other giving it the nickname, 'Yin-Yang moon', and making it seem to vanish as it rotates.

Figure 2.4.12 (right) Iapetus from *Cassini-Huygens* showing its walnut ridged shape and (left) the extreme reflectivity contrast across it
(NASA *Cassini-Huygens* Mission; www.jpl.nasa.gov/spaceimages/details.php?id=PIA12604; www.jpl.nasa.gov/spaceimages/?search=iapetus&category=#submit)

Dumfounerin Munes

Unruly bairns the Munes,
Whaes planetary mithers dae their best,
Tae keep thaim aa in some celestial order;
There's Mrs Jupiter wi aa her muckle brood,
Saxty-seeven satellites rinnin widd,
The fowr biggest; breengin, bruisin bruits!
– Io, Ganymede, Europa an Callisto.
Claucht bi Gallileo's telescope,
They hained awa their saicrets fir a while;
Tortured Io, pu'd an raxed ilk day,
Her face aa streekit wi sulphurous tears.
Europa an Callisto's cranreuch faces,
Aiblins harbour oceans,
Salome veiled,
Aneath layer upo layer kilometres thick o ice,
That cuid gie bield tae extra-terrestrial life.
Ganymede birls, muckle as Mercury,
Jundies his maw,
A braw, braw lad,
He fancies hissel a planet,
But gravity aye maun keep *him* in his place!
Whiles roond an roond their sibs the ithers race;
Metis, Leda, Carpo, Dio,
Kallichore, Carme an Eurydome,
An mony mae;
Gyte moths birlin round a flame,
Wha'll dance their tentless dance,
Till time hus tint their nems.

2.5 Saturn's Wonderful Rings

We already saw some beautiful images of Saturn and its rings in Section 2.1 (Figures 2.1.9–2.1.11). The era of interplanetary space probes and of giant telescopes brought the knowledge that all four Gas Giants are encircled by rings of debris (as well as by many moons) though with none of the other meagre ring systems coming anywhere near the magnificence of Saturn's. Saturn's rings, the four brightest of Jupiter's numerous moons, of the phases of Venus, and of evolving dark spots on a rotating Sun, were discovered when Galileo made the first celestial use of the newly invented telescope.

These discoveries eventually delivered death blows to the long-held dogmatic belief in a geocentric Universe populated by perfect bright points fixed in a heavenly firmament containing unchanging, unblemished spheres moving in perfect circular orbits around us (Section 6.2). Despite widespread familiarity with decades of giant telescope and space-probe images of all the planets in utterly exquisite detail, most of the public to this day still gasp in awe when given their first direct sighting of Saturn's rings through even a telescope as small as Galileo's. Given that such direct sightings look pretty much like the two-handled bowl in Galileo's sketch (Figure 2.1.8), this is a quite surprising reaction, most likely attributable to the fact that there is today a not uncommon but totally fallacious notion, sometimes even in well-respected newspapers, that planets are things not visible to the naked eye but only by means of telescopes and space-probes. This is despite the fact that five were well known to ancient Greek, Mesopotamian and other civilisations; and that the shivering virgin-stargazers who go out on a clear, cold, dark night, cricking their necks to peer through a telescope, realise that they are *not* looking at a screen or printed image but photons arriving on their retina direct from Saturn. As an Irish poet friend once put it, 'It's a case of cosmic contact – real Saturnian photons in your eyeball – like getting real kisses rather than in a letter or on a video.'

As well as being objects of great beauty, Saturn's rings are fascinating in their physical nature through even a very modest telescope. After 200 years of failure by many eminent minds, in 1857, James Clerk Maxwell established mathematically what the nature of the rings must be, by proving that, in order to survive stably, they must comprise many small bodies separately orbiting the planet, rather than being a single continuous solid or liquid annulus (Section 5.8). These many small icy rocky bodies, ranging from the size of houses down to the size of grains of sand and dust, have now been seen directly and can be thought of as a sort of 'failed moon'. Early Solar System debris like comets and small asteroids left over and captured when Saturn formed tend, on the one hand, to collide and break up and, on the other hand, to accrete gravitationally into moons. However, close to the planet, Saturn's gravitational tidal force is stronger than the mutual gravity which pulls them together, so they can't accumulate into a moon. It is sometimes thought that rings originate by a moon being torn apart tidally but, once fully formed by accretion, a moon is held together by the strength of its rock which can far exceed gravitational binding.

Since Saturn's rotational equator and its ring plane are inclined to the ecliptic (orbital plane of the planets), as Earth and Saturn orbit the Sun, the rings are then seen at varying angles with a pattern repeating on a roughly 30-year cycle. Twice each cycle, the Earth passes across the ring plane as seen from the Earth and the rings almost vanish, showing them to be very thin. This is clear in Figure 2.5.1 (top), which shows an HST image of the ring plane crossing. The large moon to the left is Titan, with its large shadow on the lower hemisphere of Saturn, while the other four moons, left to right, are Mimas, Tethys, Janus, and Enceladus.

The incredible thinness of the rings is even more prominent in the close-up view from the *Cassini* probe (Figure 2.5.2). They are in fact only around 10m thick, or one ten millionth of the distance across the brightest rings. This Figure also shows a number of other interesting features. In particular, the very fine radial structure of the rings (almost like a vinyl music album), which we saw in Figure 2.1.11, is seen again here in the very dark (but not quite black) shadow cast by the rings on the planet. Note also the difference in colour (yellow and blue) between the lower and upper hemisphere. Like red sunset versus blue overhead skies, this arises because of the wavelength (colour) and angular dependence of light scattering off of molecules. Lowering of the

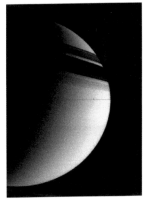

Figure 2.5.1 (left) Saturn ring plane crossing – upper image
(Erich Karkoschka, U. Arizona LPL, NASA HST; apod.nasa.gov/apod/image/saturn_24Apr_hst_big.gif)

Figure 2.5.2 (right) Saturn rings seen edge-on from NASA/ESA *Cassini* probe
(Cassini Imaging Team SSL JPL ESA NASA apod.nasa.gov/apod/ap140413.html)

temperature by the ring shadows can change the cloud and molecular composition of the cooler hemisphere. This difference is hard to see from Earth because the rings hide from us a large part of cooler hemisphere.

The reasons for the intricate spatial structure and for the subtle time variations in Saturn's rings are quite complex due to the interplay of the gravitational fields of Saturn, of its moons, and of the ring material itself. For example, the countless gaps in the ring system are at least in part attributable to resonances between the orbital periods of pairs (or more) of satellites which create places in space where the gradient of gravity is outward (or inward) in all directions, so that anything located there will tend to move away from (or stay in) there. A rough analogy would be nearby beaches along a single coastline

Figure 2.5.3 Shepherd moons Prometheus and Pandora around the F-ring (top); Daphnis generating waves in the ring material (bottom) (NASA/JPL/Space Science Institute photojournal.jpl. nasa.gov/jpeg/PIA11621.jpg ; solarsystem.nasa.gov/resources/17589/daphnis-up-close/)

having very different quantities of shells – or of flotsam – on them because of local variations in the strength and direction of waves and tidal flows relative to the shore direction. Another (extreme) example is the Great Pacific Garbage Patch, where huge quantities of human waste have accumulated around an oceanic flow node (stagnation point). A very significant role in creating and maintaining ring structure is that played by even small moons. In Figure 2.5.3, we show in the top panel the pair of small (under 100 km) 'shepherding moons', Prometheus and Pandora, which orbit inside and outside the thin F-ring, their gravitation sweeping the gaps clear and the ring filled with small debris. The black wedge at the bottom is Saturn itself. In the bottom panel, we show Daphnis (smaller than Everest) moving in the dark ring gap, which it keeps clear by gravitational sweeping of material seen as waves along the ring edge.

Aliens wha played,
Voyager's golden record,
Danced tae Chuck Berry!

Finally, one of the most thought-provoking images from any space-probe was Carl Sagan's famous *Pale Blue Dot*, showing the Earth almost 10 billion km behind Saturn as seen by NASA's *Voyager 1* on Valentine's Day 1990. The first spacecraft destined to leave the Solar System entirely was *Pioneer*, followed later by *Voyagers 1* and 2, one carrying

Figure 2.5.4 Pale Blue Dot Earth/Moon seen by Cassini (left) and *MESSENGER* (right)
(NASA/JPL-Caltech/SSI, NASA/JHU APL/Carnegie Institution DC)

a Golden Record with sounds of Earth, including music, as well as animal and other terrestrial sounds. The major annual Bluedot Festival of science, arts and music held at the Jodrell Bank site of the Lovell Radio Telescope site is named after this visionary image.

In Figure 2.5.4 we show more recent versions of the iconic *Pale Blue Dot* image concept taken by the *Cassini* (left) and *MESSENGER* spacecraft (right) on 19 July 2013. This *Cassini* image captures nearby Saturn's rings and planet Earth (bottom right) at 1.44 billion km. High magnification only just shows the Moon as a faint protrusion off Earth's right side in the *MESSENGER* image. One hundred million kilometres distant, Earth and the Moon are easily distinguished star-like features. This image was part of a campaign to search for natural satellites of Mercury.

Legends in oor minds;
Frae six billion mile awa –
A 'Little Blue Dot'

2.6 Exoplanets

Exoplanets (i) – The Hunt

My first encounter with exoplanet astronomy was in September 1969, when I was on a two-week trip around the US after a summer job in Harvard-Smithsonian Center for Astrophysics in Cambridge, Massachusetts. The trip was mainly for leisure, including jazz in New Orleans (Dixieland at Preservation Hall) and in Hollywood, LA (Miles Davis on 21 September in Shelly's Manne-Hole); Carlsbad Caverns; and the Grand Canyon. However, one study-related pilgrimage trip was to the Mount Palomar telescope. I was privileged to make that pilgrimage on the Observatory shuttle bus, and even more so to make it in the company of Dutch astronomer Peter van de Kamp, who was ascending the mountain for an observing run as part of his research portfolio. One topic in that portfolio, which was to make him (in)famous around that time, was his claim to have made the first discovery of two exoplanets – ie planets outside our Solar System – orbiting a star other than our Sun. Specifically, he eventually claimed detection of two planets orbiting Barnard's star, a near neighbour of our Sun. His claims were widely met with disbelief, if not derision, in the scientific

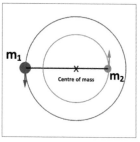

Figure 2.6.1 Binary orbit
(JC Brown)

community on the grounds that no one else's observations replicated what he claimed to see, which they attributed to faulty data analysis. Van de Kamp's doubters in the end proved to be right but, as in other scientific cases, faulty science and misplaced zeal helped drive others not just to challenge experimental results but to think about what might be done to improve the experiment. Such creative thinking ultimately helped spawn the huge explosion in real exoplanet discoveries which started in 1992, driven by advances in telescope and computer technology. In the same way, the erroneous claims of Joseph Weber (University of Maryland) in 1969 that he had directly detected gravitational waves helped to drive such research to its eventual success by the Laser Interferometer Gravitational-Wave Observatory (LIGO) team in 2016 (see Section 4.3).

So how were exoplanets eventually detected? Planets are bodies much smaller and less massive than their parent stars and, unless orbiting it very closely, are also much cooler, so they are much fainter than the star in both reflected and emitted radiation. This makes them very hard to detect directly against the bright star and most detection techniques rely on seeing some influence the planet exercises on its parent star. The first two techniques rely on the fact that planets don't really just orbit their star but rather they orbit each other about a common centre of mass. The situation is like that of an athletic hammer thrower (star) and the hammer (planet) in the fast spinning pre-release phase of a hammer throw. The thrower and hammer are turning at the same rate in revolutions per minute but, because the thrower is much heavier, their speed and their to and fro spatial movement are much smaller than those of the hammer head. In the case of the star and planet, gravity replaces the hammer handle or wire. Another analogy is a couple doing an Eightsome Reel with a hefty dancer birling a slim partner. So even if the planet/hammer/slim dancer is too small to be seen directly, we may be able to infer their presence (and their mass) by observing how far the star/thrower/hefty dancer is displaced or at what speed they are moving.

Van de Kamp was trying the former (astrometric) technique, seeking to measure the small angular movement of the star due to the invisible orbiting planet. This is most feasible for heavy planets and large star-planet separations, making the orbital period and data sampling time very long and not really feasible with van de Kamp's equipment, although this is now being achieved by the high precision of the ESA Gaia astrometric spacecraft. The second approach (spectroscopic binary) is to look for the speed of motion induced in the star (thrower/hefty dancer) by the planet (hammer/slim dancer), using the observed Spectroscopic Doppler shift in the same way as traffic police radar works. This is most viable for edge-on orbits and heavy planets close to their star, so that the observed speed is a maximum. One more exotic detection technique is gravitational microlensing. In rare cases, when a star's motion in space takes it very close to the line from us to a distant star, the gravity of the first star focusses the light from the second and makes it temporarily brighter. If the first star has planets, their gravity adds to the lensing and complicates the observed light curve of the brightening in a way that carries information about the position and mass of the planet.

However, the most productive exoplanet detection technique of all is the planetary transit/occultation method for planets in near edge-on orbits that take them across (transit) the disk of the star, blocking out a small fraction of its light from the observer. This fraction would be roughly 1 per cent and 0.01 per cent respectively for Jupiter-sized and Earth-sized planets occulting a Sun-like star, easily detectable with modern telescope technology and data processing – a fact that has brought planet discovery within reach of technically-minded amateurs. The transit/occultation process is lucidly illustrated by Lynette Cook's 1999 artwork *Exoplanet Transit of Star HD209458*, which has been reproduced with her generous permission in Figure 2.6.2. This piece was published worldwide along with the 1999 announcement of the first exoplanet discovery by its transit, previously detected only via the star's orbital wobble. Of the exoplanets now known (around 3,700 in August 2018) the majority have been

Figure 2.6.2 *Exoplanet Transit of Star HD209458* – artist's impression
(Lynette Cook extrasolar.spaceart.org/ 1999, acrylic, gouache, coloured pencil on board)

found by NASA's *Kepler* mission and the rest by an assortment of other space missions and ground-based observers, including a few by amateurs.

The *very* first confirmed exoplanet discovery (1992) was in the case of the rather bizarre situation of a pair of planets orbiting a pulsar (see Section 3.4) and, to date, only two other cases of pulsar planets are known, despite the fact that the very precise period of pulsar pulses makes it easy to Doppler track their orbits and detect small companions. Pulsar planet discoveries offer another observational saga of error driving future success, as in the case of van de Kamp and Joseph Weber. Andrew Lyne of Jodrell Bank made the first announcement of pulsar planet discovery in 1991 but withdrew his conclusion quite soon after, to be followed within a year by the Wolszczan and Frail discovery. In the case of ordinary stars, because large objects near their parent stars are the most easily detectable, many of the early exoplanet discoveries were in this category and are called Hot Jupiters. Although their preponderance is mainly an observability selection effect, the discovery of giant exoplanets very near these stars was in sharp contrast to the Solar System situation of Jupiter and the other outer Gas Giants.

The first exoplanet found (1995) orbiting a Main Sequence star (one in its hydrogen burning phase – see Section 3.4) was a giant planet in a very close fast (four day) orbit around the star 51 Pegasi, which (at 50 LY) is quite nearby on stellar scales. In 1996, the first long-period planet was discovered, a Cold Jupiter orbiting the star 47 Ursa Majoris at 2.11 AU, while 1999 saw the first discovery around a Main Sequence star of a multiple-planetary system, comprising three planets, all Jupiter-like, at widely spread distances (making them Hot, Warm and Cold Jupiters). As measurements became more and more precise, further properties like the size and density of exoplanets began to be found. For example, in 2005, the Spitzer Space Telescope made the first direct detection of light emitted by an exoplanet, the dawn of studies of exoplanetary atmospheric temperature and structure, while mid-2017 brought hints of possible detection of an exomoon. Such advances have enabled speculative but ever better impressions by artists like Lynette Cook of what it could look like on many exoplanets. Another of Lynette Cook's vast range of superb general space-art work (www.spaceart.org) is shown at the end of this subsection.

Exoplanets (ii) – The Amazing Discoveries

As a result of such progress, it became clear that the diversity among exoplanets was even more amazing than that unveiled by spacecraft in the late 20th century among planets and minor bodies of our own Solar System and vastly more complex to understand than the stars. This reflects the fact that the physics of condensed matter (solids and liquids), which constitute much of planetary masses, is very complicated compared to that of gases which make up stars (and planetary atmospheres). Below we give some examples of the extremes of various properties of exoplanets discovered to date. However, we also look at progress in the Holy Grail of exoplanet research, namely the hunt for Earth-like planets, with mass and gravity similar to Earth's, especially for those located in the so-called Goldilocks Zone of their stars. This zone is the range of distances of planets from their star where the temperature maintained by starlight is not so cold that water freezes and not so hot so that water boils but, like Goldilocks' porridge, is *just right* for life as we know it on Earth, where liquid water plays a key role in the emergence and survival of life. In addition, if life as we know it is to thrive, the planetary gravity should be strong enough to prevent thermal motion of the atmospheric atoms from leaking away into space. As mentioned in Section 2.4, the Goldilocks Zone argument for viable planets can be much too restrictive, since planets and especially moons can be kept warm and viable by geothermal/tidal energy even when remote from their central star.

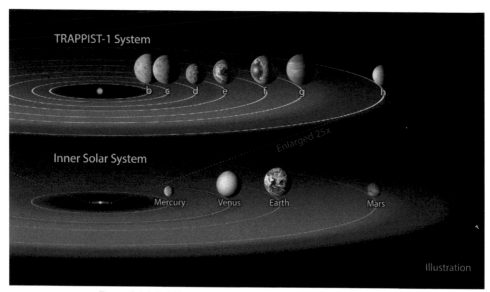

Figure 2.6.3 Goldilocks (habitable) zone for Sun and for Trappist-1
(commons.wikimedia.org/wiki/File:PIA21424_-_The_TRAPPIST-1_Habitable_Zone.jpg)

Extremes of Exoplanet Sizes and Masses

The smallest well-established exoplanet mass value is Gliese 581e at the mass of twice Earth's. Kepler-37b may be less massive (if it has Moon-like density) and is certainly the smallest yet at the radius of 0.3 times Earth's. The largest mass listed by NASA as a possible exoplanet is DENIS-P J082303.1-491201 b at around 28.5 Jupiter's mass, while the largest radius listed is HD 100546 b at around 6.90 times Jupiter's radius. These planets' masses put them very close to being very low mass brown dwarf stars rather than planets.

Extremes of Exoplanet Temperature

Kepler-70b is likely a dense/rocky exoplanet with a mass of 0.44 Earth masses and a radius of 0.76 Earth radii, orbiting every 5.76 hours very close to the surface a sub-dwarf star. This proximity must cause its surface temperature to be several thousand kelvin, which may be hot enough to melt at least some of the rocky materials. By contrast, OGLE-2005-BLG-390Lb, with a mass of around 5.5 Earths, has a 10-year orbital period, like that of our outer Solar System planets, and an estimated temperature of only about 50 K. It is one of the few exoplanets found by the microlensing method.

Extremes of Exoplanet Age

PSR B1620-26b (whose unofficial nickname is Methuselah) is located in the globular star cluster, M4, about 5,600 LY from Earth in the constellation of Scorpius, and is the oldest known extrasolar planet to date. It is one of only three planets known to orbit around a binary star, one component of which is a pulsar (neutron star) and the other a white dwarf. The planet has a mass twice that of Jupiter, and is estimated to be 12.7 billion years old, almost three times the age of Earth. At the other extreme is the newborn exoplanet K2-33b, with an estimated age of only 5–10 million years, which is about 1,000 times younger than the Sun. It has a mass of around that of Neptune and whips around its star every five days so is nearly ten times closer to its star than Mercury is to our Sun, making it hot. This situation compounds the theoretical puzzle of how so many massive planets get to be so close to their stars. Most theories involve very slow migration inward, an idea which cannot work for the very youthful K2-33b, which has not yet had much time to migrate.

Extremes of Exoplanet Remoteness

SWEEPS-04 and SWEEPS-11 are currently the farthest known exoplanets, both orbiting a star in the constellation of Sagittarius. They were detected when they and their parent star caused brightening of a distant background star by gravitational microlensing. That distant star (which bears the romantic name of SWEEPS J175853.92–291120.6!) lies approximately 28,000 LY away from us (about a quarter of the radius of our Milky Way Galaxy).

Extremes of Exoplanet Proximity and Habitability

In October 2012, the discovery was announced that the B component of the closest star-system to the Sun, Alpha Centauri, had an Earth-like exoplanet, Alpha Centauri Bb. This was seen as a major landmark in exoplanet research, since it would encourage the feeling that Earth-like planets may be common if our nearest neighbour has one. However, the claim was later refuted and attributed to a subtle error in the data analysis. At that time, this result left the closest exoplanet considered as confirmed by NASA to be Epsilon Eridani b, only 10.5 LY away from our Solar System, while the closest known rocky planet was Gliese 674b, only 14.8 LY away. Earth-like (rocky) exoplanets in their Goldilocks Zone include: Wolf 1061c, located about 13.8 LY away; Gliese

832c, 16 LY away from Earth and orbiting every 36 days within the habitable zone of a cool red dwarf star, Gliese 832; Gliese 667Cc, a super-Earth being 3.9 times more massive, which orbits red dwarf Gliese 667C every 28 days 22 LY away in Scorpius. However, the Alpha Centauri saga re-opened in August 2016 with the announcement by the European Southern Observatory that Proxima Centauri (aka Alpha Centauri C) – the third, and nearest, component of the triple star system, Alpha Centauri – has a planet. This star is a small, cool red M dwarf, only half as hot as the Sun and seven times smaller. However, its 1.3 Earth-mass planet, Proxima Centauri b, orbits it in 11.2 days at a distance 20 times closer than Earth is to the Sun. This combination of stellar size and temperature should give the planet a temperature comparable with Earth's, and put it in the Goldilocks Zone.

Early 2017 brought the amazing discovery of seven temperate terrestrial planets orbiting the star, TRAPPIST-1e, 39.5 LY away in Aquarius. This is the largest number detected in any exoplanet system, all seven orbits being closely spaced and less distant (0.01 to 0.06 AU) from their very cool star than Mercury is from the Sun. No less than three of them are in their star's Goldilocks Zone. These Proxima Centauri and TRAPPIST-1 Goldilocks Zone planets would seem like very good candidates for life to emerge. However, an important proviso is that such cool dwarf stars have strong magnetic flare activity whose radiation could be hazardous to life. This is especially true for close proximity to the star which may allow its stellar wind to blow away any protective planetary atmosphere, as is thought to have happened to Mars.

Despite these cautionary considerations, a NASA estimate based on *Kepler* data up until 2013 suggested the Milky Way could contain 40 billion Earth- and super-Earth-sized planets orbiting in the Goldilocks Zones of Sun-like stars and red dwarfs. If even 0.25 per cent (1 in 400) of these were outside the stellar flare and wind danger regimes, and heavy enough for their gravity to retain an atmosphere, that leaves 100 million life-viable stars in our Milky Way Galaxy with a mean spacing of about 10 LY. This estimate is based on the starlight-heated Goldilocks Zone meaning of life-viability. It does not allow for tidally driven heating like that on Io, which can power life geothermally and life forms like occur in Black Smokers (hot volcanic ocean bed vents) let alone life forms in some of the almost unimaginable environs of some science fiction. Examples of these are Robert Forward's *Dragon's Egg* creatures which live under the incredibly extreme surface conditions of neutron stars (Section 3.4), and Arthur C Clarke's *Out from the Sun*, where human inhabitants from a base on Mercury experience thinking activity among the electrical processes in a passing solar plasma cloud (Coronal Mass Ejection – Section 3.1). On Earth, the most remarkable life-forms in terms of viability in, and survival through, extreme conditions are the sub-mm sized

Tardigrades, colloquially known as water bears, or moss piglets (www.wikipedia.org/wiki/Tardigrade).

However, it is important to understand that when we use phrases like 'only 10 LY away' and 'next-door neighbours in space', you should not be reaching out to your travel agents to book your holiday there, nor expecting little green men first-footing you. In everyday terms, the distance to even the nearest star is 4.5 LY or 4×10^{13} km away. So, if it could operate in a vacuum and carry enough fuel etc, a jet airliner would take about 6 million years to get there. Thus, if you don't want to hibernate and would prefer to get there and back in a lifetime, you will have to get your skates on and speed up to at least half the speed of light. Unfortunately, the whole of humankind is unlikely to offer you all the energy they consume in a year, which is roughly what is needed to get just your body (with no fuel, air, food or drink) to $0.5c$ (Section 1.1). So, interstellar

Figure 2.6.4 (left) NASA JPL High g Planet Poster of HD 40307 g for daring skydivers
(NASA JPL)

Figure 2.6.5 (right) NASA JPL Planet Red Grass Poster of Kepler 186f where grass is tuned to photosynthesis the starlight
(NASA JPL)

travel is unlikely for the foreseeable future. However, on a brighter note, since light travels at the speed of light, it might be feasible for us to exchange to and fro light or radio signals with the nearest star up to ten times in a human lifetime, if anyone is there and willing to listen to our crazy communications. Despite these practicalities, NASA JPL created of a set of fun Exoplanet Tourist Destination art posters some of which we show here. These include:

- Kepler 16b – where you are never lonely, as its double star gives you two shadows
- (Figure 2.6.4) At 40 LY, HD 40307 g is one of a six planet system that has a 200-day year, and lies in the Goldilocks Zone of life-comfort temperature. Its high surface gravity (twice that of Earth's) should retain an atmosphere and give sky-diving tourists a big thrill.
- (Figure 2.6.5) Kepler 186f – where grass is always redder for maximum photo-synthesis, using the light from its Red Dwarf star.
- (Figure 2.6.7) 50 LY-distant Trappist 1E offers the best (seven) exoplanet tour
- PSD J318.5-22 – offering endless (but chilly) nightlife as it has escaped from its star (see Appendix Figure A4)

Some more extreme and bizarre destinations could include:

A Diamond Exoplanet?

55 Cancri e is an 8-Earth mass rocky exoplanet. If made of a carbon-rich material, then a third of it might be diamond which would be worth around $2.7x 10^{31} at our current diamond market price per carot. However, even ignoring the 40 LY transport costs, flooding diamond markets with three Earth-masses of diamonds would make them worth far less than salt.

A Jet-Black Exoplanet?

Kepler 1b, 750 LY from Earth, was identified in 2011 as the darkest known exoplanet, reflecting less than 1 per cent of any light that hits it. The planet's mass and radius indicate that it is a Gas Giant with a bulk composition similar to that of Jupiter but, like many exoplanets, it is located very close to its star and so belongs to the Hot Jupiter class.

Doomed Exoplanets?

Exoplanet HD 209458b was found by the Hubble Space Telescope in 2004 to exhibit strong carbon (c) and oxygen (o) absorption features in its UV spectrum. These are interpreted to be as a result of intense evaporation by the powerful starlight of c and o molecules from what may be 'dead' cores of former Gas Giants whose outer H and He layers have already been fully evaporated. Exoplanet WASP-18b has a mass of ten Jupiters, close to the lower mass limit of brown dwarf stars needed for gravity to initiate nuclear fusion in its core and become a star (Section 3.4). So, it just qualifies as an extrasolar planet, with an orbital period of less than one day, skimming very close to its host star. The resulting tidal forces are expected to make its orbit a death spiral, eventually merging with the star in under a million years.

The artistic potential of the exotic world of exoplanets is huge, as is exemplified by Lynette Cook's cosmic artwork portfolio (available at www.extrasolar.spaceart.org),

Figure 2.6.6 Exoplanet HD 222582d and moon; this artist's impression by Lynette Cook is from the surface of a moon, with the (possibly) ringed planet in the background; acrylic, gouache, and coloured pencil on board, 2000
(Lynette Cook – www.nonsolarspaceart.org)

one more beautiful example of which we show here in Figure 2.6.6. Of this exoplanet
HD 222582 b and its moon, Lynette says:

> HD 222582 b (of over seven Jupiter masses), orbiting a 137 LY distant G3 (Sun-
> like) star HD 222582, has a highly eccentric 572-day orbit taking it from 0.39 AU
> to 2.31 AU from the star. The water on this world's satellite, if one exists, goes
> through seasonal periods of melting and refreezing in the warmth from the star.
> This view is from the surface of a moon, with the (possibly) ringed planet in the
> background.

Figure 2.6.7 NASA JPL's Tourist Poster of seven exoplanet
system Trappist1E
(NASA JPL www.jpl.nasa.gov/visions-of-the-future/)

Goldilocks an the Three Exoplanets

Goldilocks hud taen the strunts,
Earth no fit tae leeve in,
Ah'm gangin oan an astral hunt,
Oot amang the heivins,

This braw blue pearl that aince oor pride,
Nou's gane tae potterlowe!
Ah'll fin anither place tae bide,
In some hyne cosmic howe.

Be't bears or gods ah ne'er wis sweir,
Tae venture intil nicht,
Sauf bield we'll win, the best o gear!
Too hoat? Too cauld?? Juist richt!!!

The road wis lang bi EasyJet;
Six million year or mair!
Tae rax the nearest planet's yett,
The journey lang an sair.

'This wullna dae', cries Goldilocks,
An shouts oan Albert Einstein,
'Ah need a spaceship, ane that rocks!
Twa years is ma deidline!'

Al chowed his pencil near tae spails,
But foun hou tae warp space,
An syne Wee Goldilocks set sail,
Fareweel the human race!

She skited doun a big black hole,
At Sagittarius,
An near-haund licht speeds she'd tae thole,
Abuird her starry bus.

She passed the time bi spinnin yarns,
O ferlies seen ilk day,
A hunner million life gien starns,
Bide in *oor* Milky Way.

Juist finnin yin tae dae the joab?!
She'd kill fir e'en a glisk!
Oot there thair shairly ae sib globe;
Too hoat? Too cauld?? Juist richt!!!

Kepler-70b? nae guid!
Whaur rocks melt wi the Sun,
The pairritch frae its plate juist slid,
Birslin oan the grund!

Hoth, o icy Star Wars fame,
(Wha's rael nem's faur owre lang!)
Its aamaist *'absolute zero'* claim,
Keeps aathing in a jam.

Minus mair's twa hunner degrees,
Pairritch disnae dae,
A cog o cranreuch glacial debris?
No whit ye waant tae hae!

Oor heroine hauds forrit tho,
Thair shairly ither Earths,
Fir that bricht starn owre there we'll row,
In by Aquarius.

The 'Goldilocks Zone'! we'll hae a luik,
Aye, that yin, Trappist-1,
Mind, strang magnetic flares tae jouk;
Thae stellar winds that burn!

Syne doun oan fremmit saunds her ship,
Haes laundit by the shore,
The hinnied air sae sweet, she slypes,
Doun frae the cabin door.

Twa munes that blink amang the starns,
Embers ruby licht;
Pairritch hotterin in the pan,
Too hoat? Too cauld?? Juist richt!

CHAPTER 3
The Sun, the Stars and Nebulae

OUR SKY IS teeming with stars, as we can see in the Hubble Space Telescope image in Figure 3.0.1, which covers only a tiny angle of the sky equal to that of a one pence piece seen from 20 m. Looking like fairy lights, this is in fact a globular star cluster (Section 3.3) called NGC1898, located in the Large Magellanic Cloud, which is a small galaxy (Section 4.1) orbiting our own Milky Way Galaxy. In this chapter, we will mainly be talking about the Sun and relatively nearby stars in the Milky Way, but the nature and physics of stars is similar across the cosmos.

Figure 3.0 Celestial fairy lights
(ESA/NASA/Hubble (apod.nasa.gov/apod/ap181003.html))

3.1 Solar Facts, Figures and Phenomena

That big bricht yin's oors,
Thae wee yins skinkle at nicht;
But oot there, oot there...

In this section, we mainly explain how the Sun, as we normally see it, works as a fairly typical, very long-lived, very steadily self-luminous star, but also address in some detail the observations and workings of its very dynamic and complex outer layers (its atmosphere) and their impact on the Solar System. These are sharply revealed by viewing the Sun in spectral bands (wavelengths) well removed from the visible light in which we see the Sun's bright photosphere. To set the scene, Figures 3.1.1 and 3.1.2 illustrate respectively how different the Sun looks even across a fairly modest range of ultraviolet (UV) wavelengths and how its UV appearance changes over its 22-year activity cycle (solar 'climate'), as well as on the much shorter timescales (solar 'weather') we discuss later. The creative visualisation of the Sun in Figure 3.1.1 from visible to extreme ultraviolet wavelengths

Figure 3.1.1 SDO's Multi-wavelength Sun
(NASA/SDO GSFC Scientific Visualisation Studio https://apod.nasa.gov/rjn/apod/ap131221.html)

uses image data from NASA's orbiting
Solar Dynamics Observatory (SDO). The
lower half shows the yellowish Sun in
broad band visible light, while the upper
segments show it in narrow wavelength
bands ranging (left to right) from visible
light (the human eye wavelength range)
to extreme UV. Each is dominated by
spectrum lines emitted mainly in specific
temperature ranges, thus picking out successively hotter parts of the Sun's atmosphere. However, before getting further
into the fascinating science of the Sun's
long-term life as star and of its variability
in the shorter term, we touch briefly on a

Figure 3.1.2 Evolving solar activity in EUV light
(NASA, SDO https://www.nasa.gov/mission_pages/sunearth/news/gallery/
Max_cycle2.html#)

couple of the countless examples worldwide of the Sun in human legends.

The Sun in Legend

Our nearest star, the Sun, has borne diverse names throughout history and around the
world, and has long been the stuff of worship, myth and legend, both as the bringer of
light and warmth, but also in its diurnal appearance and disappearance. Such worship

Figure 3.1.3 Some Sun Legends
Left: NNW America coast legend of the raven stealing the Sun
(artwork Nathan Jackson – wikipedia.org/wiki/Nathan_Jackson, photography Shelly Stallings, Ketchican Museums ref KM 87.2.11.1)
Right: Chinese legend of the dog Tiangou eating the eclipsed sun
(commons.wikimedia.org/wiki/Category:Tiangou#/media/File:Zhangxian02.jpg)

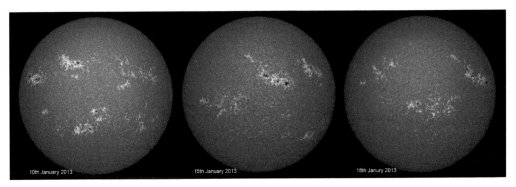

Figure 3.1.4 Narrow waveband visual image sequence of the Sun with dynamic features revealing evolving magnetic activity as they rotate across the Sun
(R Arnold, Isle of Skye, www.robertarnold.co.uk/photographs/astro/Solar/index.htm)

is hardly surprising, even in today's hard-nosed science-, technology- and money-driven world, given that the Sun's radiation and gravity are what enables life and that the constituent atoms of our planet and our bodies came from the Sun as it formed. Among the legends of indigenous peoples of the American northwest are those from the northern northwest coast people – Tlingit, Haida and Tsimshian. These legends include the Raven as both hero and trickster, such as his theft of the Sun to bring light and heat to the Earth. Figure 3.1.3 (left) shows a superb painted carving of this legend by renowned artist Nathan Jackson from that region (Raven Side of the Tlingit Clan) displayed in the Ketchikan Alaskan Heritage Centre. The small image on its right in Figure 3.1.3 depicts the very different Chinese legend of the dog Tiangou eating the Sun during an eclipse.

In Section 3.4, we will explain where the Sun fits into the stellar scheme of things, what stars are and how they are born, live and die, and, in Section 2.1, we discussed how planets form as part of star formation. Because we are so close to it, we can see the Sun's surface in exquisite detail which makes solar physics a vast subject, parts of it quite messy. We do not attempt to cover even a small fraction of that subject, but just to give some basic interesting solar facts and figures. Even in normal visible light, the Sun shows itself to undergo Sunspot activity and to rotate, facts known to Galileo and shown in Figure 3.1.4 by amateur astrophotographer Robert Arnold (Waternish, Isle of Skye), a frequent contributor of images to Spaceweather solar image archive.

Solar Vital Statistics

In the grand stellar scheme, our Sun is a fairly small yellowish dwarf star, far smaller than the giants among the top end of stellar sizes, but bigger than the average size of all stars that is weighted toward the huge number which are smaller than the Sun. In terms of spatial location, it is nothing special either, lying well out in the cosmic suburbia of the Milky Way Galaxy's disk (100–1,000 billion stars). However, in terms of planetary and human scales, it is mind-blowing and immensely important in many ways, including:

- It is about 100 times the Earth's size and a million times its volume and mass.
- It rotates in around 25 days (but nearer 35 near the poles) – see Figure 3.1.4 – as first realised by Galileo from his daily drawings of Sunspots as they traversed the disk.
- Being about 270,000 times closer than the nearest star, it appears 270,000 times larger and 73 billion times brighter in the sky than it would at the distance of that star.
- In absolute terms, the total solar nuclear fusion power and its total light output is 4×10^{26} w, equivalent to 1 billion 100 Megaton TNT-equivalent bombs per second.
- This core power is released as nuclear gamma/x-ray photons, but these take over 100,000 years to emerge at the surface 0.7 million km above. This is because, instead of 'flying' straight out of the Sun's surface in the 2.2 sec of light travel time, they scatter off particles every millimetre or so, their travel is impeded and they drift outward in a slow random stagger. This is a bit like trying to leave a crowded bar or football stadium. Each is also downgraded into thousands of mainly visible photons before finally escaping at the photosphere and flying across the Solar System vacuum to reach Earth in just 8 minutes. In some stars, including the Sun, some of the heat is carried part-way outward by convection and some by radiation. This convection process sets up streams of rising hot and falling cooler gas, like above a home radiator, which cause a honeycomb-like pattern of convection cells on the Sun's visible surface (the photosphere) – see Figure 3.1.6. Convection plays a key role, along with rotation, in creation and dissipation of the Sun's magnetic features.
- The chain of nuclear fusion processes long-believed to power the Sun's radiance also generates neutrinos at a huge rate. These ghostly chargeless particles interact very weakly with matter and escape directly from the Sun's core to arrive at its surface. Measurement of their arrival rate using huge detectors, deep underground to prevent cosmic rays interfering, provides a direct test of the solar fusion theory. This test worryingly contradicted predictions for many years

29th March 2006. Antalya, Turkey. Prof John Brown

Figure 3.1.5 Solar Eclipse Antalya March 2006 – composite image blend by Robert Arnold of 1/200 to 1 sec exposures taken by JC Brown with hand-held digital SLR with 135 mm lens
(JC Brown and Robert Arnold)

after the first detections in the 1960s, the measured rate being about 1/3 of that predicted. It was ultimately shown in 2002 that the solution to this major solar neutrino problem was not to be found in the solar fusion theory, but in a nuclear process called neutrino oscillations by which 2/3 of the solar fusion neutrinos convert into undetectable forms in flight.

- Of the 4×10^{26} w of total solar power output steadily arriving at Earth-distance, 1 AU or about 150 million km from the Sun, the area of the Earth intercepts about 4×10^{18} w. If this could all be collected and utilised, it would provide around 10,000 times mankind's current power consumption (almost 50 per cent of which is by China and the US).

- This amounts to around 1.4 kw per m^2, about the area of a person, which is firmly in the life-favouring Goldilocks Zone (Section 2.6) for the Sun – obviously, or we would not be alive to write this stuff.

- The orbital planes of the Moon around the Earth and of the Earth around the

Sun cross at two points. If a full Moon is at one of these points when the Sun is at the other, the Moon is eclipsed in the Earth's shadow, though it has an eerie red glow due to sunlight refracted and scattered through the Earth's atmosphere (Figure 2.3.8). However, if the (new) Moon and the Sun arrive at the *same* crossing point, then the Moon's shadow eclipses the Sun. A lucky coincidence is that, although the Moon is around 400 times smaller than the Sun, it is also around 400 times nearer, so the Moon's disk almost exactly hides the Sun's as seen from a narrow path on Earth (total or annular eclipse). Elsewhere on the Earth, the eclipse is partial. During total eclipses (which last a maximum of 7.5 minutes), the Sun's brilliant photosphere vanishes and its faint but magnificent outer (coronal) layers are revealed to the naked eye. This creates one of Earth's most awesome sights, striking fear into the hearts of the uninformed as if our life-giving Sun were being consumed (Figure 3.1.5)

Detailed Appearance of the Sun

From the astronomer's perspective, the Sun's proximity offers an unprecedented chance to study a 'typical' star in exquisite detail. We can distinguish very small features on its surface (around 70 km for a large solar telescope). The huge flux of photons from the Sun also allows study of it in narrow wavebands from radio to gamma-rays (corresponding to different plasma temperatures/energies) and at high spectral and time resolution. The dark Sunspots on the Sun's large, bright optical disk were first recorded by early Chinese observers and are now known to be one of many types of magnetically controlled complex structures in the solar atmosphere on both small and large scales and at all levels in the atmosphere, far more complex than the smooth bright (5,800 K) yellow ball we see in white light (see especially Figures 3.1.6–3.1.7). The former is one of the highest ever resolution optical images of a Sunspot region revealing the intricacy of plasma structure

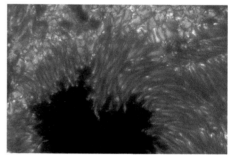

Figure 3.1.6 Sunspot close-up
(Royal Swedish Academy's Solar Telescope on La Palma apod.nasa.gov/apod/ap110918.html)

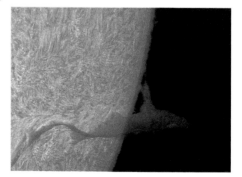

Figure 3.1.7 Solar prominence in H-alpha light
Prominences are dense cool regions in the tenuous hot corona and supported by magnetic forces
(Robert Arnold, Isle of Skye)

Figure 3.1.8 Erupting
prominence in XUV seen
by SDO with other sites of
energy release
(NASA GSFC AIA Team apod.nasa.gov/
apod/ap180916.html)

at photospheric level created by the magnetic field and by the rise and fall of convection cells like those above a radiator.

Understanding Solar Atmospheric Structure and Dynamics

Explaining the complexities of solar atmospheric structure, largely governed by magnetic fields, has improved greatly since the advent of ever more sophisticated space-borne instrumentation, which enables study of the corona other than during brief eclipses of the photosphere by the Moon. These instruments include white light corona-graphs viewing sunlight scattered from coronal electrons, and high energy instruments viewing UV, x-rays and Gamma-rays. The latter are emitted by the hot corona and by fast particles generated in it during stormy magnetic weather events like flares, filament eruption and coronal mass ejections (CMEs). Today, multiple spacecraft also allow stereo imaging of the Sun which, together with spectroscopy and magneto-grams, help inference of the 3D distribution and evolution of the atmosphere, vastly more complex than the mostly smooth visible light disk. Ongoing puzzles regarding the physics of the solar atmosphere include: the coronal heating problem; the speed of magnetic energy release in events like prominence eruption, CMEs and flares; and the efficient acceleration of highly energetic electrons and ions in such events.

The Coronal Heating Problem

Besides the intricacies of coronal structure, the average temperature (around 2 million K) of the corona as a whole is in itself a long-standing puzzle. Above the hot photosphere, just as below it, one would expect the temperature to continue to drop off as distance from the Sun increases. (The second law of thermodynamics indicates that heat always flows from hot to cold.) So, something other than light must be delivering solar energy into the corona. Dissipation via electric currents of magnetic energy is widely believed to be what heats the corona, although getting energy out of the magnetic field quickly enough is problematic for material with such low electrical resistance. Most of the Sun, and especially its corona and core, are so hot that they are not just in the gaseous state but in the plasma (4th) state of matter, meaning that enough of their atoms have lost electrons to endow the gas with sufficient charge (and, hence, electric currents) for it to acquire often dominant electrical and magnetic properties. The heat of the corona also drives the solar wind, which carries away about 1.5 million tonnes of solar material per second into interplanetary space at around 500 km/sec. While this may sound like a lot, when it reaches Earth, the wind material has been spread out to just one

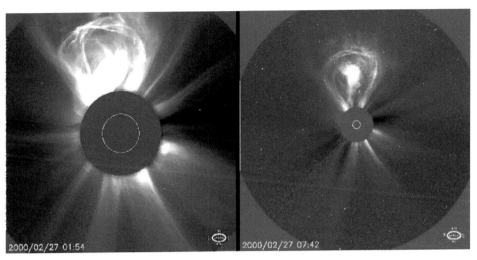

Figure 3.1.9 Coronal mass ejection (CME) sequence
(NASA SOHO LASCO)

proton per cm³. Furthermore, the solar wind mass loss rate amounts to only about 1/3 of the Sun's mass loss rate in the form of sunlight. The power needed to drive the wind flow, and that to heat the corona, are both tiny compared to the Sun's visible (photospheric) light output. As the late Sir Patrick Moore (presenter of the BBC's *The Sky at Night* for 55 years, a TV record) always emphasised, 'The corona may have a high temperature but there's very little heat in it.' This is because the 2 megakelvin corona is so tenuous that, if you were in it, its heat flux on your skin would feel merely cosy, although you would be vaporised by intense sunlight from the much cooler but much denser 5,800 K photosphere below.

Solar and Space Weather

Like coronal heating, the other major physics puzzles of the solar atmosphere also involve the problem of fast enough magnetic energy conversion, although in transients, rather than semi-continuously. These are solar flares – loosely speaking, very powerful, impulsive localised deposition of magnetic energy into plasma heating and sometimes an outward blast of material. CMEs can involve as much mass and energy as flares but occur somewhat more gradually (with less power), with magnetic expulsion of a large coronal mass into interplanetary space. Both often also give rise to acceleration of energetic electrons and ions partly because the changing magnetic field (like a dynamo) generates a large-scale high voltage drop and partly because shock waves from the explosion generate lots of small-scale voltage drops.

Solar Probes

The most recent daring adventures in exploring the Sun are NASA's *Parker Solar Probe*, launched in August 2018, and the eagerly awaited ESA *Solar Orbiter* (2020). These missions have complementary goals, but both are designed to fly far closer to the Sun than any predecessor. *Parker* will, by means of a seven-year series of orbital slingshots around Venus, reach an elliptical orbit swinging it to within about 6.4 million km (9 solar radii) of the Sun's surface, far closer than Mercury's orbit and almost to a point where it is in the static outer corona rather than the outflowing solar wind. The technology of these probes is very sophisticated in order to protect the delicate instruments from destruction from the intense sunlight, like the wax-fixed wings of the Icarus legend.

Sun-skimming Comets as Solar Atmospheric Probes

The fact is that we already have valuable data on the inner solar atmosphere from supersonic probes passing 100–1,000 times closer to the Sun's surface than the *Parker* or *Solar Orbiter* or any robot probe of ours ever will. These probes are the nuclei of very close Sun-grazing comets, notably Comets Lovejoy 2011 and ISON 2012 – quite blunt instruments with no electronics or orbit controls, but nevertheless quite hardy and very valuable probes of the inner solar atmosphere. Formerly, Sun-grazing comets were only visible near the Sun in white light coronagraphs showing photospheric light scattered by the electrons of the comet, in the same way that the corona is visible in coronagraphs and during eclipses (Figure 3.1.10). This technique is limited to the minimum radius where coronagraph masks can hide the photospheric light sufficiently, usually more than about 1.5 solar radii from the Sun's centre.

However, at the UV imaging wavelengths covered by SDO instruments, the quite cool photosphere is faint and it is potentially possible to see UV emission from any very close Sun-skimming comets, which

Figure 3.1.10 Comet Lovejoy composite coronagram
(NASA SOHO archive)

are large enough to survive until they reach there. This requires a fairly large comet (around 1km) to avoid total evaporation in the intense sunlight there. (It is sunlight – and tidal forces – and *not* the tiny effect of the tenuous but hot corona heat flux that vaporises and fragments Sun-skimming comets. This is contrary to the media's and even NASA's persistence in deluded expressions of amazement at relatively tiny Sun-skimming comets surviving the high coronal temperature.)

The first such Sun-skimming comet survivor and the second seen by SDO UV telescopes low in the Sun's atmosphere was Comet Lovejoy 2011, whose orbit took it just 0.2 solar radii above the Sun's surface on 16 December 2011, travelling hypersonically at 536 km/sec. Contrary to most predictions, it was seen to re-emerge from the Sun, although very substantially dissipated. Ingress and egress (entry into and emergence from the inner solar atmosphere) are shown in Figure 3.1.11, which clearly displays how the comet probes the atmosphere for us. In particular, note the bend in the comet tail during ingress (left image) caused by the Sun's magnetic field; and, in both ingress and egress (right image), and notice the repeating light/dark striations across the comet tail. These have been interpreted as arising where the comet nucleus encounters denser regions of the Sun's atmosphere, which heats the vaporising comet matter enough to ionise (become electrically charged) and become redirected along the Sun's local magnetic field. Thus, UV movies of such Sun-skimming comets not only tell us about the properties of the comet but also about the density and magnetic field structure of the Sun's atmosphere.

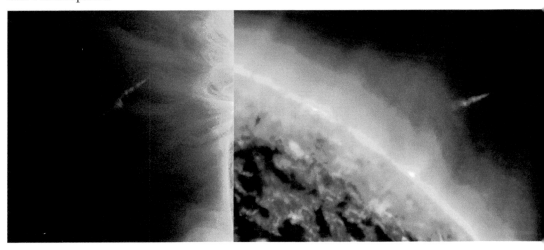

Figure 3.1.11 Solar atmosphere ingress and egress of Comet Lovejoy as seen in UV by SDO
(JC Brown collage / NASA/Lockheed SDO AIA Team and collaborators)

Geomagnetic Storms and Aurorae

All this magnetic activity means that, while life is enabled by the Sun's steady light (and by its gravity keeping us in orbit), space weather created by solar magnetic activity – solar flares, CMES and fast wind streams from coronal holes – can have adverse effects on the Earth, especially for humans. This is even more so in the case of smaller dwarf stars – even less massive, cooler and more magnetised than the Sun – where flaring activity is very powerful and threatening to life in the Goldilocks Zone discussed in Section 2.6. Even in the case of the Sun, when a mass of high-speed solar plasma collides with the Earth's magnetosphere, the shaking magnetic field acts like a huge dynamo creating high voltages. These accelerate strong currents and fast particles into our atmosphere, making it glow and flicker like a fluorescent light tube – creating the beautiful aurorae discussed below. The downside is that they can sometimes disrupt power lines, transformers and telecommunications, and can also be a radiation hazard to spacecraft and high-altitude flights. To paraphrase what CP Scott said of the word television, 'These words are half-Latin and half-Greek. No good will come of them'. One might say something similar about the aurorae Borealis and Australis. Because of the economic and social effects of these Sun-induced geomagnetic storms, there is a lot of investment worldwide in studying space weather, such as in the US's Living with a Star research programme, in hopes of learning to predict and prepare for bad solar weather events.

Finally, we turn to the beauty and wonder of aurorae Borealis (northern) and Australis (southern). Because aurorae arise from fast particles descending magnetic field lines near the planetary poles, they are most readily and most often seen from high latitudes but it is a common myth that one can only ever see them under very dark skies and at high latitudes. In fact, aurorae are not entirely unknown down to near the equator and they are not uncommonly seen from around Glasgow ($56°$ N), sometimes even amid city light pollution. Figures 3.1.12 and 3.1.13 show respectively a selection of such Scottish auroral views, as well as views looking overhead from Norway and looking down from ISS. A very fine view of the aurora in Scotland is that on the cover of this book. It was taken by amateur photographer Tadhg Macleod (of Kyleakin) at Ashaig, Isle of Skye, and is remarkable for its showing at its left end a fine example of a Strong Thermal Emission Velocity Enhancement (STEVE). Such STEVE events are distinct from but related to the aurora and are not yet well understood, having only become widely reported recently, presumably partly because of advances in observing equipment.

CMES and fast solar wind streams from solar activity also act on other planets with magnetic fields – most notably Jupiter and Saturn – to generate energetic particles, auroral activity and radio emission from them (Figure 3.1.14). Interestingly, the Jovian auroral light shows strong spectral signatures of elements like sulphur ejected by the volcanoes of its moon, Io.

Figure 3.1.12 Selection of auroral images from Scotland
Top left to bottom right:

Callanish, Lewis
(Emma Rennie www.callanishdigitaldesigns.smugmug.com/)

Doune
(Douglas Cooper)

North Connell
(David Banks www.davidbanksastro.com)

Glasgow Maryhill
(Iain Hannah)

Figure 3.1.13 (left) Aurorae from Norway and from the ISS over Pacific NW America
(Johnny Henriksen, Harstadt, Norway, via www.Spaceweather.com
NASA/ESA ISS [Astronauts Scott Kelly/Tim Peake])

Figure 3.14 (right) Aurorae of Jupiter and Saturn
(NASA/ESA/HST www.nasa.gov/feature/goddard/2016/hubble-captures-vivid-auroras-in-jupiter-s-atmosphere
NASA, ESA, Hubble, OPAL Program, J. DePasquale [STScI], L. Lamy [Obs. Paris])

Solaris

Enow tae gar wir eident, auncient forebeirs
Bigg Stonehenge, Stenness or Ring o Brodgar.

The Pueblo's gowden mean at Mesa Verde;
Teotihuacán's Pyramid o the Sun.

Thon auld sang 'ken thy sel' o Delphic fawm,
Cairven oan the Temple o Apollo.

We maun airt oot the haund o Amon-Ra,
Mangst the muckle Pyramids o Giza.

Aboriginal Gnowee's ruffie lowes,
As she hunts ower the yird fir her loast wean.

Whiles Chinese lore, the magic dug o heiven,
Is flegged frae bitin aff a daud o Sun;

The weans aa screich an yaw an dunner gongs,
Tae lowse the jowler's jaws frae thon eclipse.

Aruna, Surya's whipman, hoys his cairt,
Its seven naigs drive oan at daw or gloamin;

Sól, o Nordic myth, bade aince the samen,
Alsvid an Arvak yerk him throu the lift;

An haunds that raxed intil the Trundholm glaur,
Syne brocht yer solar cairtie tae the licht.

E'en thon praicious seeds athin yer nieve,
Hained frae time-worn tombs an gien some watter,

Solaris face micht shine oan their descendants,
When they ae day rax roond faur fremmit orbs.

3.2 Polaris

If you asked the average person in the street to name a star, they would most likely name the Pole Star (Polaris). If asked what the brightest star in the sky is, apart from the Sun, many, if not most, would again say Polaris, although in fact it is only the 50th brightest star in the sky. The brightest is in fact Sirius, the Dog Star, in the constellation of Canis Major, which follows Orion the Hunter low in the sky (at European latitudes), with Sirius' bright blue light twinkling strongly after passing through the atmosphere.

So why is such a faint star as Polaris *so* famous? One could say that Polaris is famous, like many top celebrities, simply because they are in the right place at the right time. Polaris happens at the moment to lie very close to the point in the sky directly above the North Pole of the Earth – ie along the axis about which the Earth spins west to east once a day. As a result, we see all celestial objects (Sun, Moon and stars etc) revolve around us once a day east to west around the point in the sky along the Earth's axis – the north celestial pole. Some rise and set but those lying far enough north in the sky never set but describe 'circumpolar' paths around the pole or trails in long exposure photographs (Figures 3.2.1 and 3.2.2). Polaris lies very close to that point, revolving in such a small circle that it almost looks static.

Figure 3.2.1 (left) Circumpolar star trails from the University of Glasgow quads
(Hugo Cheung, 4th year neuroscience student from Hong Kong)

Figure 3.2.2 (right) Circumpolar star trails with aurora from North Connell;
here, Polaris lies just out of the frame best to show the auroral display on the north horizon
(David Banks, Oban)

Anywhere in the Northern Hemisphere, the height of the Pole Star above the north point of the horizon equals the latitude of the observer. Furthermore, the positions of the constellations in their circling around it can tell us the time, like hands on a clock face, conversely our longitude if we know the time. So, Polaris and the stars around it have long played a key role in navigation and, throughout history, have featured widely in literature worldwide. It also appears on the Alaskan Flag, thanks to the creativity of 13-year-old Alutiiq orphan, Benny Benson, who submitted the unanimously judged winning entry out of many hundreds to a competition to design a flag for the (then) Territory of Alaska in 1926. In his own words:

The blue field is for the Alaskan sky and the Forget-Me-Not an Alaskan flower (now the State Flower). The north star is for Alaska [the most Northerly State]. The Dipper is for the Great Bear, symbolising strength.

Figure 3.2.3 Alaska Flag with Polaris and Plough
(commons.wikimedia.org/wiki/File:Flag_of_Alaska.svg)

However, things have not always been, and will not always be, thus. Everyone knows that the Earth spins once a day and orbits the Sun once a year, but less well-known is that the Earth's spin axis direction wobbles like that of a spinning top completing a (precession) circuit once every 26,000 years. So over long periods of time in the past and the future, the Earth's axis was pointing nowhere near Polaris, nor any bright star – though, from time to time, it comes close to one which then becomes the north star of that era for a while. The most striking case is the bright blue-white star, Vega (Alpha Lyrae), in the constellation of Lyra the Harp (5th brightest star after the Sun and seven times brighter than Polaris). Vega occupied the north star site around 14,000 years ago and will do so again 12,000 years from now.

But, right now, the position is nearly occupied by the moderately dull Polaris at the end of the tail of the Ursa Minoris (Little Bear) constellation which has a shape similar to, but much smaller in size than, Ursa Majoris. However, while today's Polaris is pretty feeble in brightness compared with Sirius and with the earlier epoch North Pole star, Vega, it far outshines its South Pole counterpart (the brightest star near the point above Earth's South Pole). At present this is the star, Sigma Octantis (also called Polaris Australis), which is 25 times fainter than Polaris and only just visible to the naked eye.

It is slightly closer (a half degree or the angular diameter of the Moon) to the South Pole than Polaris is (2/3 of a degree) from the North.

Its location above our North Pole and on the Alaskan flag are, however, not Polaris' only claims to fame. It is also a physically interesting triple star. The brightness of Polaris as a whole is dominated by the light of the main component star, Polaris Aa. This is an old supergiant star that has become a pulsating Cepheid Variable – a class of variable star discovered by Henrietta Leavitt to have pulsation periods related to their intrinsic brightness. This discovery was crucial to our measuring of the distances of remote stars and ultimately the size of the Universe. The basic principle involved is that if you know the power (w) of a lamp, you can work out how bright it will look at different distances. So, conversely, if you measure how bright it looks, you can figure out what its distance is, making it a Standard Candle. Leavitt's discovery means that measuring the pulsation period of a star tells us its wattage (see Sections 1.1 and 4.1).

The other two components of Polaris, Polaris Ab and Polaris B, are both dwarf stars quite like the Sun and over 100 times fainter than Polaris. Polaris B is not physically connected with Polaris A and is far enough away from it across the sky to make Polaris A, B easily seen separately as a double star even in small telescopes. Polaris Ab, on the other hand, orbits Aa in about 30 years and has long been known to exist because the orbital motion of Polaris Aa shows up as periodic Doppler shifts in its spectrum. However, Ab lies so close to Aa in the sky and is so faint that it had never been seen directly until 2008, when it was recorded by the HST.

Constant

'I am constant as the northern star,
Of whose true-fix'd and resting quality
There is no fellow in the firmament.'
—*Julius Caesar* (iii, i, 60–62)

Ask ony body tae nem a starn,
An like as no ye'll fuin,
They ettle oan thon mirk wee scad,
Polaris.

Dwaiblie as a caundle flame,
Hingin frae Ursa Minor's tail,
This shilpit craitur shuid hae nae infit,
An yet...

We coost oor een in sairch o constancy.

Be't luver's hairts,
Or Scotrail trains,
Or friens wha bide aye true,
Human beings fix oan aa that's leal.

Lik ye, Polaris –
No the brichtest,
No the biggest,
No the maist kenspeckle.

Still an oan, yer constant licht hus bin oor guide,
Ower ocean, desert, jungle or icy waste;

Sma-boukit, slicht, ye bide North Star,
Bringin thaim we lo'e aye saufly hame.

3.3 The Pleiades and Other Star Clusters

Figure 3.3.1 (left) Pleiades open cluster
(Graham Malcolm, Kilbarchan)

Figure 3.3.2 (right) Pleiades location
Pleiades location photo with Glendalough round tower, March 2015 (top right)
(Rob Hurson, CC BY-SA 4.0, Wikimedia Commons)
Pleiades locator chart – a straight line from Orion's belt takes you to the Pleiades (bottom right)
(Randy Culp www.rocketmime.com/astronomy/)

The Pleiades (Figure 3.3.1) is an open star cluster in the constellation of Taurus the Bull. It is a bright, compact group in a part of the sky north of Aldebaran (the red eye of Taurus) with relatively few bright stars around it, so it is quite eye-catching in autumn and winter skies.

Figure 3.3.2 shows the location of the Pleiades, first as seen with Glendalough's round tower, photographed by Rob Hurson in March 2015, the cluster lying just left of centre in the sky to the right of the tower. Also visible are Sirius (bottom left) and the Orion constellation between it and the tower, the Hyades cluster being hidden here by the tower. The elegant 30 m tower in County Wicklow, Ireland, was built by monks almost 1,000 years ago, twice as long ago as the starlight seen here was emitted by the Pleiades. Below this fine photo, we have placed a Pleiades locator chart created by Randy Culp.

The Pleiades and the double cluster in Perseus – see below – are among the most superb celestial objects to view through small telescopes and even through modest binoculars. Its shape somewhat resembles a small version

Figure 3.3.3 Logo of the Subaru Corporation and 2019 Subaru XV (Subaru (UK) Ltd)

of the Plough (Big Dipper), which is part of the Ursa Major (Great Bear) constellation. The Pleiades is also commonly named the Seven Sisters, after the number of stars visible in it with good naked eye vision. Most casual viewers see just the five brightest stars (Alcyone, Atlas, Electra, Maia, and Merope) but some with acute vision under ideal conditions have reported up to 20.

The Japanese name of the Pleiades is Subaru (which means 'unite') and a schematic of the six brightest stars in that cluster forms the commercial logo of Subaru Vehicles (Figure 3.3.3).

The Pleiades is among the brightest-looking of star clusters. It contains a total of some 3,000 stars moving together through space as a gravitationally-bound group. It is about 13 LY across and lies about 444 LY from Earth. Such open clusters are mainly young groups of stars formed together by gravitational shrinkage and consequent heating of clumps in a large gas cloud (Sections 2.5 and 3.4) moving together in galactic disks. Its age is estimated at around 100 million years – about 1/50th of the Sun's. The cluster is also associated with a fine example of reflection nebulae, seen as (mostly blue) light reflected/scattered from

Figure 3.3.4 The Seven Sisters by Elthu Vedder, 1885
wikipedia/commons/f/f8/The Pleiades%28 Elihu Vedder%29.jpg)

Figure 3.3.5 Perseus Double Cluster
(Graham Malcolm, Kilbarchan)

an interstellar cloud of cold gas and dust particles through which the cluster is passing.

The Seven Sisters also feature widely in early history, mythology, art and also in agriculture and sailing, their autumn return to the evening sky being a valuable calendrical marker. In fact, they were the subject of myth and legend in almost every culture on the planet (Figure 3.3.4). For example, Greek legend has it that Orion (who first appears as a great hunter in Homer's *Odyssey*) pursues the Seven Sisters, so Zeus transforms them first into doves, then into stars to comfort their father. On the other hand, in Cherokee myth, they are seven wayward boys defying their parents, while the Aboriginal version is of seven beautiful ice maidens whose affections toward men were as cold as the stream from which they came.

Between the Pleiades and Orion in the sky, also in Taurus, lies the Hyades (closely linked to the Pleiades in mythology), the open cluster nearest the Earth (153 LY), made up of hundreds of stars with an estimated age of 625 million years. The brightest stars form the v-shaped head of Taurus the Bull, whose eye is the bright red star, Aldebaran. Aldebaran does not belong to the gravitationally bound Hyades cluster itself but lies at just 65 LY along the line of sight to that group. Two other famous naked eye open clusters are the Praesepe (or Beehive) cluster in Cancer and the magnificent Double Cluster (h and χ) in Perseus which was described in an early sky observer's guide as looking like 'diamonds on black velvet' (Figure 3.3.5).

A totally different class of star clusters from the Opens is the Globulars. These comprise many more and much older stars than open clusters, and are tightly bound by gravity into dense spherical collections which orbit at large distances from us (100,000 LY) in the halos of their galaxies. Large galaxies typically have hundreds of Globulars, each with hundreds to millions of stars. The brightest Globular is Omega Centauri in the Southern Sky. M13 in the north – Hercules – is another bright, frequently imaged object. It lies at a distance of around 22,000 LY, is about 100 LY across, and about 10 billion years old, a hundred times older than the Pleiades. It is estimated to contain several hundred thousand stars. Figure 3.3.6 shows M13 imaged by Scottish amateur, David Banks, whilst Figure 3.3.7 shows the HST view of the stellar density in its heart.

Figure 3.3.6 (left) Amateur image of Globular Cluster M13 in Hercules
(David Banks, Oban)

Figure 3.3.7 (right) Hubble ST image of heart of Globular Hercules Cluster M13
(ESA/NASA HST, www.spacetelescope.org/images/opo0840a/)

Subaru

Doc pu's up in his fantoosh motor-caur,
The latest wird in Japanese technology;
His Forester, shure fuitit as a mountain goat.
Proodly he wipes some glaur frae its starry badge.

'Subaru' ah say, 'The Seeven Sisters',
An we digress tae crack anent cosmic things…

The Pleiades, a peerie version o The Pleuch,
That glent heich up in the hairst-time lift,
Jist abune the widd rid ee o Taurus,
Joukin the horns o that radgey auld buhl!

Mair's twa hunner million years hae passed,
Syne they wir cast abreid,
Tae soom in timeless heivins.

Aiblins orbs o distant gas an fire,
Else pawns dernit frae Orion's ire,
Wha Zeus himsel fixt in their place.

Whiles Neolithic fairmers nae dout traced,
At Ness o Brodgar winter's dowie oncome;
Yer twinklin ferlies whuspert saftly tae thaim,
Tae hain the seeds that handsel in the Voar.

Vedder's neo-classical erotica,
Hus ye'se dancin tae yer maister's tune;
Owre yer heids in gowden leashes abune,
Yer wunnerfae starns birl an blink fir aye.

Doc, suitably enlichtent, kennles ye up,
An guns the willin growl o that flat-four.
Myth an legend melded intil ane;
Excalibur or racing at Le Mans.

When they are celestial dust,
The Sisters wull shine oan.

Star Clusters

Shoals o silver darlins
Dairtin throu the lift,

Gowpinfou's o diamonds,
A Sultan's waddin gift!

Tinsel hanked oan branches,
Oan ilka Christmas tree,

Sequins shewn oan 'Strictly'
That blinter oan TV.

Polisht chrome oan Harleys,
That thunder throu ilk toon,

Tin cans tied tae bumpers,
Oan auld farrant honeymoons.

Rhinestones at the Opry,
Silver coins in fountains,

Keekin oot an aeroplane,
At snaw capped winter mountains.

Swarovski crystal craiturs,
Oan gless shelves o dazzlin licht,

Fireflies that flauchter,
An set lowe the tropic nicht.

3.4 Birth, Life and Death of Stars, and their Associated Nebulae

Stellar Birth in Cool Gas Clouds

The life stories of stars form a wonderful tale involving some of the most beautiful nebular gas cloud objects in oor big braw cosmos, and their corpses form some of its most amazing objects – white dwarfs, neutron stars and stellar black holes. They are also directly linked to the origins of planets (Section 2.5) and to heavy chemical elements (sometimes termed metals), hence to life itself. In Figure 3.4.1 (see also frontispiece image of Horsehead Nebula) we show images all taken by Scottish amateur astronomers. They show diverse sites in our vicinity of the Milky Way Galaxy, where stars are forming from huge cool molecular gas clouds. Figures 3.4.2 and 3.4.3 are close-up views into the Eagle and Horsehead Nebulae, respectively taken with the NASA and ESA HST.

We saw in Section 1.3 that, well after the Big Bang, when the expanding hot plasma Universe cools and becomes transparent, and gravity starts to dominate over gas pressure in dense and large enough gas clouds, these collapse and form many of the cosmic structures we see today. Among these are protostars with a wide range of masses, which form from clumps as seen in the Horsehead and Eagle close-ups from HST. Note that the very dark regions in visible light (like the Horse's Head itself) are not *holes* in the emitting nebular material but *silhouettes* of colder absorbing foreground material. All of these shrinking objects grow hotter due to release of their

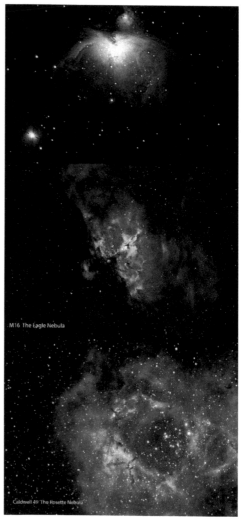

M16 The Eagle Nebula

Caldwell 49 The Rosette Nebula

Figure 3.4.1 Stellar birthplaces
(top to bottom)

Great Nebula in Orion
(Alex Houston, Alloa)

Eagle Nebula
(Charlie Gleed, Edinburgh, IMM 1951–2017)

Rosette Nebula
(Charlie Gleed, Edinburgh, IMM 1951–2017)

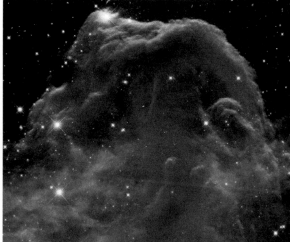

Figure 3.4.2 Pillars of Creation close-up view
inside the Eagle Nebula
(NASA/ESA HST Heritage www.nasa.gov/feature/goddard/2017/
messier-16-the-eagle-nebula)

Figure 3.4.3 Horsehead Nebula close-up in dust-
penetrating Infra-red light
(NASA/ ESA HST Heritage heritage.stsci.edu/2013/12/original.html)

gravitational energy as they fall (like in a hydro power station) and they glow with increasing brightness.

In the late 19th century, William Thomson (Lord Kelvin) and Hermann von Helmholtz claimed that such gravitational shrinkage actually powers the Sun and stars. However, shrinkage of a gas cloud down to the Sun's present size could only power the Sun's light output for about 30 million years, while the fossil evidence shows that the Sun had been shining for around 100 times longer than that. At that time, Kelvin would not heed evidence from mere geologists but later research by others (including Arthur Eddington) showed that nuclear energy could power the Sun for 100 times longer than gravity, in line with the fossil evidence. The primary nuclear process involved has to be by fusion of hydrogen into helium (like in a hydrogen bomb), since the Sun is mainly hydrogen. Nuclear fission (radioactivity and atomic bomb) energy needs *very* heavy elements. These are very rare, created by fusion in exploding stars and in coalescing neutron stars.

Protostars and Fusion Onset

Hydrogen fusion can only stop cloud shrinkage and replace the gravitational energy lost by radiation if and when the cloud's core gets hot and dense enough for fusion to work – the colliding nuclei need to defeat the repulsion from their positive charges.

This conversion from a protostar to a true star can only happen for gas clouds above a minimum mass. Protostars below that never become true stars – with long fusion enabled lives – but simply glow by gravitational shrinkage alone until high density quantum pressure sets in so that they stop shrinking and fade away as Brown Dwarfs similar to the White Dwarf corpses of low mass true stars after fusion supply runs out, described below. A common misconception is that protostars only start shining when fusion starts. In reality, they shine as soon as they start shrinking and are warmed by gravitational energy. What fusion does, if and when triggered by this heating, is to enable them to shine for a much longer time.

Stellar Populations and Minimum Masses

In the early PBB (pure hydrogen) Universe, the minimum protostar mass for fusion to occur has been shown theoretically to be much larger than a solar mass. The massive true stars of that first-generation epoch are contrarily called Population III and remain hypothetical rather than definitely observed. This is possibly because any of them young enough still to be seen shining would be very remote and so very faint. By contrast, the current highly evolved Universe contains the whole range of chemical elements in the Periodic Table (though some in very small quantities). These have been created by fusion in the cores of earlier stars and released across the Universe when these blow out most of their mass as winds or explosions in their death throes (see overleaf). It turns out that such traces of heavy elements act as powerful catalysts for nuclear fusion and enable it in protostars of much lower mass than those of Population III, namely

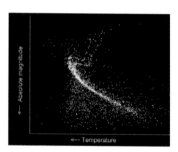

Figure 3.4.4 HR Diagram relating stellar luminosity L to colour (surface temperature) for two star clusters of different ages (yellow and blue)
(WikiCommons - User: Worldtraveller)

about 0.1 Suns or 100 Jupiters. Most stars like the Sun and their contemporaries in spiral galaxy disks (Section 4) belong to a third generation of stars – called Population I – whose material include some ejected during the death throes of second-generation stars – called Population II – which formed from material including that left over from Population III. Population II stars are mainly found in elliptical galaxies and in the extensive haloes around the dominant disks of spiral galaxies.

Stellar Maturity/Main-Sequence

When heavy enough protostars reach fusion onset, they enter stellar adulthood the longest phase of their luminous existence. Stars in this phase belong to the stellar Main Sequence (MS) because, in Hertzsprung-Russell (HR) diagrams showing brightness versus temperature (colour), they lie along a clear curve at positions defined by their masses, the most massive having the biggest, hottest (bluest) and brightest surfaces (Figure 3.4.4). Though the heaviest stars have much more hydrogen to fuse than lighter ones, their rate of fusion and radiation are very much higher, and they consume their H-fusion energy in a much shorter time so they 'live fast and die young'! For instance, the Sun's MS lifetime is around 10 billion years while, for a 20 solar-mass star, it is only a few million years. This MS life ends when hydrogen fusion in the stellar centre has created a substantial core of helium (which, being much heavier, sinks), at the surface of which the temperature/density may be too low for hydrogen fusion. How things evolve after that is quite complex and depends on the mass of the star. Because hydrogen fusion heating is declining, the tendency is for the core to shrink and heat again. In all but the least massive stars, this initiates fusion of core helium into carbon with some hydrogen burning in a shell above that. The onset of helium burning is a profound stage in the stellar story, stellar nuclear alchemy transforming helium into carbon, the Rosetta Stone of life as we know it and non-existent in the early Big Bang Universe.

The Death Throes of Low Mass Stars

This internal restructuring redistributes energy and causes the outer layers of moderate mass stars (around 0.5–8 Sun masses) to expand and cool to form Red Giants – a few times cooler than the Sun but hundreds of times larger. When the Sun reaches this stage billions of years hence, it will engulf the Earth, though life as we know it will be gone long before then (Section 8.1) – probably even including the hardiest beasts known to us, the Tardigrades. The low surface gravity of these low mass stars allows light and gas pressure to drive off their outer layers as a Red Giant wind which does two amazing things. The first is to eject heavy elements like carbon, nitrogen, and oxygen created by interior fusion into the mainly hydrogen material of interstellar space where it can become part of new protostars, thus of new stars and planets and of any life which forms there. This is the origination of the saying, 'we are made of star stuff'. The second is that collision of the wind with interstellar gas creates hot dense luminous shells called Planetary Nebula (PN). The name arose because, in the simplest cases, this interaction creates a roughly spherical bubble looking like a planetary disk – eg the Ring

Nebula in Lyra and the Blue Snowball in the Andromeda Constellation (see Section 7.4). However, due to complicating effects like rotation of the star and wind, the presence of magnetic fields, and supersonic flow turbulence, many PNs have vastly more complex and beautiful shapes (Figure 3.4.5). Another less well-known but simpler PN example is NGC 7662 – the Blue Snowball – shown later in Figure 7.4.4 – since it is one of the objects the authors observed on their SDO Observatory visit with New Cumnock Burns Club.

White Dwarf Formation

Deep inside these low mass stars, the energy-depleted but still very hot core shrinks under the overlying weight until it becomes *so* dense (around 1 tonne per cm³) that a new type of pressure – quantum electron degeneracy pressure – sets in. This pressure does not depend on temperature, only density, so can balance the weight while allowing the core to cool off as a dead White Dwarf (WD) star, onto which the residual outer material settles. These objects are around Earth size (10,000 km),

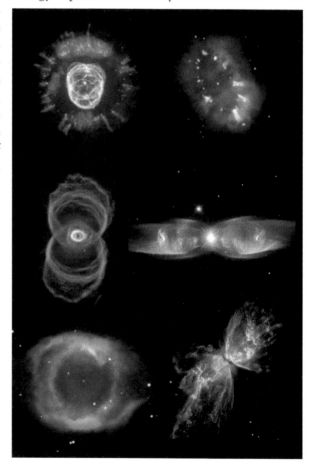

Figure 3.4.5 Planetary nebulae –
diversity of structure
Top left to bottom right:
(i) Eskimo
(NASA/ESA HST: A Fruchter and ERO Team (S Baggett,
Z Levay STScI), R Hook (ST-ECF))
(ii) Pearl Necklace
(NASA/ESA HST and Hubble Heritage Team (STScI_AURA))
(iii) Hour Glass
(NASA/ESA HST: R Sahai & J Trauger (JPL) and WFPC2
science team)
(iv) Twin Jet
(NASA/ESA HST (Judy Schmidt))
(v) Helix
(Charlie Gleed, Edinburgh (IMM 1951–2017), from his own
Coonabarabran robotic telescope)
(vi) Butterfly
(NASA/ESA HST and Hubble SM4 ERO Team)
Grid production
Bob King, *Duluth News Tribune* (Facebook: Astro-Bobs-
Astronomy-for-Everyone)

curiously the most massive ones being the smallest because the quantum pressure needed to balance their larger weight demands a disproportionately higher density and smaller size. WDs radiate at initial temperatures exceeding 100,000 K, which affects the emission from and colour of their PNs – eg the vividly blue Blue Snowball. One example of a White Dwarf stellar corpse is the binary star companion Sirius B of Sirius A, which (excluding the Sun) is the brightest star in the sky and 10,000 times brighter than Sirius B. Curiously, the ancient mythology of the tribal Dogon people of West Africa allegedly depicts Sirius as having a companion star.

The Explosive Deaths of Massive Stars

The post MS evolution and fate of massive stars is very different from lower mass ones since their weight generates much higher central pressures and temperatures. The fusion sequence hydrogen to helium to carbon then continues creating an 'onion skin' stellar structure, with the heaviest elements lying deepest and each fusion stage lasting shorter and shorter times until iron (Fe) is reached. At that point, suddenly the nuclear 'fuel tank' is empty and we reach the Iron Catastrophe because fusion of iron and heavier nuclei does not *release* energy but *requires* it. The natural tendency of heavy elements is to break up by radioactive nuclear fission. The loss of an energy source to maintain pressure leads to very rapid collapse of the overlying mass, aided by easy escape of neutrinos generated during neutron formation and by onset of a relativity effect which limits electron degeneracy pressure. This collapse only stops, if at all, when the core is so small (around 10 km) that it comprises mostly tightly packed neutrons at a density of around 100 million tonnes per cm^3 (which equates to around the mass of all mankind in a teaspoon). The accompanying huge release of gravitational energy results in a colossal ultra-hot (supernova or SN) explosion blasting off the entire outer part of the star at about 10,000 km/sec across interstellar space as a supernova remnant (SNR), which becomes, like PNs, another beautiful form of nebula. The light energy released within hours by these exploding stars exceeds that in their entire previous lifetime and temporarily outshines their entire host galaxy. (Please note, there are in fact other types of supernova which reach equally explosive ends but via different routes, which we do not discuss here.)

Large as the SN light output is, even more SN energy goes into neutrinos. On 24 February 1987, the first fairly near SN in the few centuries since the advent of the telescope, spectroscope and modern astrophysics was seen in the Large Magellanic Cloud (LMC). It reached easy naked eye visibility in the Southern Hemisphere. It also saw the advent of non-solar neutrino astronomy with the detection of a total of 25 neutrinos worldwide – the first and thus far only SN so detected. It is estimated that the SN emitted a total of around 10^{58} neutrinos, of which around 10^{17} passed through the giant (thousands of tonnes) detectors and around 10^{15} passed through each of us in a few seconds after spreading out over their 150,000 LY journey to us. Of these 10^{17}, only 25 were detected but they delivered invaluable new information on SNs and neutrinos and gave birth to the new subject of (non-solar) neutrino astronomy. The rest of this vast number passed through the detectors like ghosts, since neutrinos interact very weakly with matter. (An interesting curiosity is that the volume containing all of the 10 sec burst of SN1987a neutrinos was a near perfect spherical surface of thickness only a fraction 10 Light sec/150,000 LY = 2×10^{-12} of its radius, or two parts in 10^{12}.)

Neutrino astronomy is now a growing discipline. For example, in September 2017, an ultrasensitive neutrino detector called IceCube, embedded in the Antarctic ice, detected a single neutrino which careful detective work showed to have coincided with events detected by many other electromagnetic-instruments, including the space-borne Fermi high energy detector. Using this blossoming Multi-messenger Astronomy approach, the event was traced to the activity of a Blazar – a supermassive black hole in the centre of a galaxy (Chapter 4) – located 3.7 billion LY away in a direction near the star Bellatrix in Orion.

Ultrahot SN explosions are vitally important to us because they drive creation by fusion of the heaviest of elements (upward from iron like nickel, lead etc) which are swept out, along with lower mass elements, in the SNR. These ultimately inseminate the cosmos and future stars and planets with these star-created elements, many of which (like iodine and germanium) play vital roles in our bodies and in our technology. SNs are by no means the sole source of heavy elements. In particular, the mass loss in strong

Figure 3.4.6 SN 1987a evolution
(Collage creation: JC Brown; SN 1987a before and after (two left panels): David Malin; SN87 remnant after three decades in context (right panel): Mosaic Hubble Heritage Team; Close-up of current remnant (top right corner): Dr C Burrows, ESA/STScI and NASA)

Figure 3.4.7 Cassiopeia A – composite Image of wavelengths from three NASA missions
(Spitzer IR – red; Hubble ST visible – yellow; Chandra X-ray – blue) NASA/JPL-Caltech/Stacie/CXC/SAO www.jpl.nasa.gov/spaceimages/details.php?id=pia03519)

stellar winds from stars like Red Giants, blue Super-giants, and Wolf-Rayet Stars are also important suppliers, while an important source of heavier elements like gold is now believed to be the coalescence of neutron stars, first detected via its pulse of gravitational waves 17 August 2017, seen by LIGO (see Section 4.3).

In viewing SN remains, we must always remember that what we see now is how the SNR was when it was younger – by a light-travel time, which depends on its distance. In fact, SN 1987a in the LMC took place about 158,200 years ago but, at the moment, what we see is how it was like just about 32 years after the explosion with the angle of the expanding shell just visible. It also remains by far the closest SN explosion seen by us since the invention of the telescope. The following lists, in increasing age of discovery date, closer SNs (in our own galaxy) spotted over the last millennium and recorded visually (except Cass A, which was missed at the time, and Cygnus Loop which appeared pre-history about 7,000 years ago).

Figure 3.4.8 Crab Nebula
(NASA, ESA, HST J. Hester, A. Loll (ASU) apod.
nasa.gov/apod/ap180909.html)

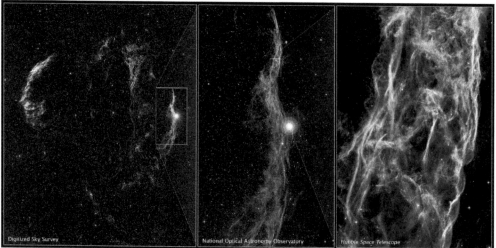

Figure 3.4.9 Cygnus Loop /Veil Nebula – optical band – zoom in sequence
(NASA, ESA, the Hubble Heritage Team (STScI/AURA), Digitized Sky Survey ((DSS))
FSTScI/AURA, Palomar/Caltech, and UKSTU/AAO), and TA Rector
(University of Alaska, Anchorage) and WIYN/NOAO/AURA/NS www.hubblesite.org/image/3621/news/3-nebulae)

- SN 1987 a (Figure 3.4.6)
  Distance ~ 160,000 LY; size ~ 0.5 LY
- Cass A ~ 1720 (Figure 3.4.7)
  Distance ~ 11,000 LY; size ~ 10 LY.
  Explosion unnoticed – now mainly a very strong radio source
- Oph 1604 (Kepler's SN)
  Distance ~ 16,300 LY; size ~ 16 LY
- Cass B 1572 (Tycho's SN) (Rough date – not recorded then; now a strong radio source)
  Distance ~ 9,000 LY; size ~ 25 LY
- Crab Nebula (Figure 3.4.8)
  Distance ~ 6,500 LY; size ~ 5 LY
- Cygnus Loop /Veil Nebula (Figure 3.4.9) ~ 5000 BC, not recorded
  Distance ~ 1,500 LY; size ~ 100 LY (10^{15}km)

The overall angular size of the Cygnus Loop is huge and about six times that of the Moon – due to the fact that it is about four times nearer and has been expanding for seven times longer than the Crab. On the other hand, the SNR of SN 1987a is still tiny in the sky, being very remote and only expanding (as seen by us) for a few decades. One of many beautiful local features in the Veil is Fleming's *Triangular Wisp* (Figure 5.9.2), named after Williamina Fleming (Section 5.9).

At the other angular extreme, in the centre of the image of Cass A (Figure 3.4.7), the SN remnant neutron star – around 10 km in size – can be seen as a sharp turquoise dot (Chandra x-ray data). Such a tiny object can only be bright enough to be seen if it is extremely hot and so emits x-rays. It also has to be young enough not to have cooled yet. (This is unless, as in the case of the Crab Nebula neutron star, it has rapid enough spin and a strong enough misaligned magnetic field to become a pulsar and emit synchrotron radiation from highly accelerated electrons – see below.)

All SN explosions seen in the telescope era, other than in 1987 in the LMC, have been in external galaxies (Section 4.1–4.3), the nearest being SN 1885a in the Andromeda Galaxy M31 (Section 4.1), which is 15 times further than the LMC satellite of our own Galaxy (Section 4.1). A much more recent but about 10 times more distant (about 25 million LY) supernova was SN 2017eaw (Figure 3.4.10), one of no less than ten in the last century discovered in spiral galaxy NGC 6946, also known as the Fireworks Galaxy. This 3.5-hour shot (featured in NASA APOD on 25 May 2017) was taken from Ao, Italy, with a SkyWatcher Newtonian 200/1,000 by a young Italian medic, Dr Paolo Demaria. A keen amateur stargazer since age ten, he is now a keen participant in a digital imaging group.

Figure 3.4.10 Supernova SN2017eaw in Fireworks Galaxy NGC 6946 is arrowed at lower left in a wide field view teaming with bright foreground Milky Way stars and with a nice Open Star Cluster (Section 3.3) NGC 6939 at 5,000 LY near top right
(Paolo Demaria GP in Cuneo, Italy, apod.nasa.gov/apod/ap170525.html)

The huge luminosity of SNs in fact makes it possible to discover them at large distances, even with modest equipment, provided it is suitably set up to monitor the sky thoroughly for changes. For example, at his Coddenham Observatory (Figure 3.4.11), Scottish amateur Tom Boles discovered a total of 155 SNs in quite distant galaxies – for example, in NGC6246 (Figure 3.4.12), about as far away as Fireworks Galaxy NGC 6946 – a world record number for one individual. It also makes them suitable Standard Candles for measuring the distances of the remotest galaxies with the largest telescopes and, hence, studying cosmological evolution – see Sections 1.3 and 4.

Neutron Stars, Pulsars and Stellar Black Holes

The core object left behind in an SN is a stable neutron star (NS), around 10 km in size but only if its mass is less than about three Suns. Though formed very hot, the small size of NS makes them hard to see directly and they are most commonly observed either via x-rays from a gravitational accretion disk of hot matter pouring inward from a nearby giant

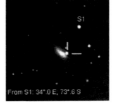
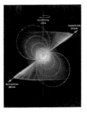

Figure 3.4.11 (left) Tom Boles at his Coddenham Observatory
(www.coddenhamobservatories.org)

Figure 3.4.12 (centre) Tom Boles' SN Discovery in NGC6246
(www.coddenhamobservatories.org)

Figure 3.4.13 (right) Pulsar model – in this artist's impression the neutron star is shown in white and its magnetic field is blue. The pulsar's radiation beams (yellow) from the north and south magnetic poles swing around the oblique rotation axis (red) and sweep past the observer like beams from a lighthouse
(B Saxton, NRAO/AUI/NSF (public.nrao.edu/gallery/parts-of-a-pulsar/))

companion star. Solitary NSs can occasionally be seen when young and very hot by their x-ray emission (see discussion of Figure 3.4.7), but more commonly via the rapid regular pulses of (mainly radio) emission neutron stars, which have formed pulsars, discovered by Jocelyn Bell in the 1960s. These amazing objects, with their ultra-precise ticking like signals from intelligent extraterrestrial life (widely termed back then as little green men) turned out to be neutron stars with very high spin rates and magnetic fields created as part of their shrinkage, and with the magnetic field axis skewed away from their rotation axis. The rapidly changing magnetic field created acts as a dynamo (Faraday), creating a high voltage and accelerating a strong current of very fast electrons. These electrons emit highly directional synchrotron radiation which sweeps the sky like a lighthouse beam as the NS rotates, and is seen as regular pulses by observers lying in the right direction (Figure 3.4.13). These several solar mass objects, shrunk to the size of a town and spinning at up to 1,000 times per second, form clocks so accurate that they are of value to us as such. In the case of the slowing binary pulsar, they also enabled indirect proof of the correctness of Einstein's predictions of gravitational emission by accelerated masses. Their material is incredibly dense – roughly the mass of the entire human race crammed into the size of a sugar cube – or a thousand billion times denser than water. This density arises because of the colossal gravity of a neutron star, so strong that the energy needed to climb up the height of a sugar cube (around 1 cm), equals that needed on Earth to climb Mount Everest around 300,000 times.

For massive star remnants of more than about three Suns, the effectiveness of neutron degeneracy pressure also becomes limited by relativity and the NS gravity becomes so high that relativistic curvature of space disconnects the remnant mass from our part of space time, and it vanishes into a black hole. Black holes are bodies with infinite

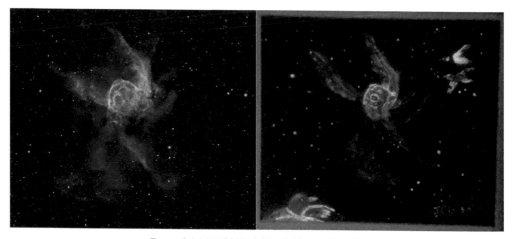

Figure 3.4.14 Wild Duck/Thor's Helmet Nebula
(Left – Feraphotography www.feraphotography.com/index.html
Right – JC Brown 2015, Acrylic on Canvas Board)

gravity from which not even light can escape and, unsurprisingly, they are the topics of endless research and of imaginative science fiction and art. Their name is possibly the technical term best known to the general populace worldwide. Usually people think of black holes as very massive things, but they can have any mass, depending on where and how they are formed. When formed from dying stars as discussed here, they have to be massive enough for their gravity to defeat neutron quantum pressure of the stellar remnant matter. However, in the very early Universe (Chapter 1), the material was so dense that Primordial black holes of very low mass could also form. On the other hand, supermassive black holes are believed to form commonly in galactic centres (Chapter 4). Though their name suggests invisibility, they are in fact detectable in a variety of ways including: indirectly through radiation from the hot accretion disk of matter falling into them (eg in the Cygnus x-1 – a binary), which also occurs for neutron stars; the light-ray bending effect (gravitational lensing) on images of the sky behind them – see Section 4; in principle, by Hawking Radiation predicted in quantum mechanics as leakage of energy across their otherwise closed event horizons – this radiation by black hole Hawking evaporation is brightest and fastest for low masses; and the distinctive form of gravitational wave pulses emitted by black hole mergers, as predicted by General Relativity. The first black hole merger detection was on 14 September 2015 and announced by LIGO and VIRGO collaborations on 11 February 2016 (see Section 4.3).

Finally, we mention another regime of nebulae generated by very massive and luminous stars – up to around 100 Suns – in a late phase of very intense rapid mass loss.

NGC 3372 The Eta Carina Nebula

Figure 3.4.15 Carina Nebula NGC 3372 & (insert) Eta Car Homunculus
(Charlie Gleed MIM 1951–2017 from his Coonabarabran robotic scope;
NASA/ESA /HST SM4 ERO Team www.jpl.nasa.gov/news/news.php?feature=7184)

Their large sizes and extremely hot surfaces (up to about 100,000 K) make such stars incredibly luminous (up to 5 million Suns), the huge radiation driving away most of their outer layers and creating turbulent hot emission nebulae with complex structures. Examples include the Thor's Helmet (or Wild Duck) Nebula NGC 2359 around 30 LY across in Canis Major (Figure 3.4.14), and the ten times larger Carina Nebula NGC 3372 (Figure 3.4.15). The latter contains a complex of stars amongst the most massive and luminous known to us, with annual wind mass loss rates of around 1/1,000th of a solar mass, or 10^{11} times the solar wind. One of them, eta Carina, underwent a massive outburst in the 1840s, creating the ten solar-mass homunculus nebula deep inside the Carina Nebula (see insert in Figure 3.4.15).

Deuks an Dugs

A Fantasia on the Thor's Helmet Nebula, AKA the Wild Duck Nebula

'Flat oot lik a deuk's fuit' says ane auld Scots saw...
But their naethin 'flat' aboot this braw ferlie!

Hotterin awa in its cloods o gas,
Awa ayont Canis Major...

Ye maist can see Sirius strain at his leash;
Aamaist as bricht as oor ain dazzlin star!

Nae wunner this deuk haes taen tae flicht,
Flegged bi thon fierce jowler i' the lift!

Some cry ye fir the Norse o Thor's winged helmet,
An aye, ah see hou thon cuid be,

Thair somethin steers the Celt in me!

Gin we hae the smeddum tae claim the heivins,
Micht we no as weel glaum the Cosmos tae?!

Tackin oor shilpit eemage oan wunners faur awa...

But mind lik ah say thon pawkie saw,
An ithers like it; 'As seeck as a dug!';

Spewin oot masses o fum an reek,
As muckle as oor Sun, ilk hunner millennia.

Aiblins wuid as cartoon Butch or Daffy,
Solar wuins breenge throu molecular cloods,

Tae flush ye frae yer leemitless cuiver...

A lowe o oxygen atoms, glentin gas,
Thon's ae description, aye, but...

Shairly thair a beauty here desairves the haund
O some Da Vinci or a Michelangelo?

Wings whirrin in a flurry o bluey-green,
A fremmit Mallard hoys in flicht throu space,

Whiles Sirius yowls an howls his thortert rage.

CHAPTER 4

Galaxies and the Large-Scale Structure of the Cosmos

Thon's the jizzen beds,
Whaur starns hae their natal hour;
Rays o licht rax oot.

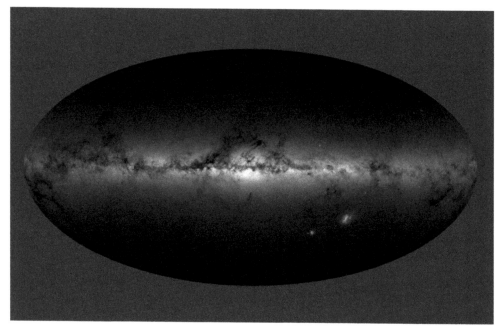

Figure 4.0.1 Milky Way Galaxy reconstruction from Gaia survey – simulation of how Milky Way looks from outside based on a reconstruction from individual measurements for nearly 1.7 billion stars by the Gaia astrometric space mission
(ESA, Gaia, DPAC apod.nasa.gov/apod/ap180427.html)

4.1 The Milky Way and Andromeda

If one looks carefully at a clear night sky, other than in today's most over-lit urban areas, one can just make out a diffuse band of light (eg to our left of Orion and passing across Cassiopeia and Perseus). This is the Milky Way, now known to be a galaxy, and usually called The Galaxy, inside which we live. It is very prominent under good dark rural skies

Figure 4.0.2 Milky Way over Scotland – Callanish, Lewis (top)
(Emma Rennie callanishdigitaldesigns.smugmug.com)
Close-up over Coll Dark Sky Isle (bottom)
(Kathleen Moore, Edinburgh)

anywhere (see Figure 4.0.2) and, contrary to popular belief, it is visible (though faintly) in moderately lit urban areas. In early human epochs, it must have been a very obvious feature of the night skies across the whole planet as seen in Figure 4.1.1. This image shows a zoom-in series of Milky Way shots taken by Scottish amateur astronomer, Mark Hill (of Neilston and Loch Doon), who used the same modest astrophotography kit he uses to great effect at home to take these photographs on two dream nights in Namibia. From top to bottom these are:

- Wide angle shot showing the diffuse light emitted by our Milky Way Galaxy's billions of stars spread out over 100,000 LY from us and, to its left, the Zodiacal Light formed by the scattering of light by dust near our Solar System plain just tens of Light Min away.
- Zoomed-in shot showing the seemingly countless stars that make the Milky Way resemble a river of light.
- Close zoom-in showing one of the many dark features in the Milky Way caused by absorbing dust and gas between us and the stars. The random shapes of many such features give rise to names such as in this case Leaping Horse.

The Milky Way bears different names in different cultures including Silver River in China and South East Asia, River of Heaven in Japan, and Winter Street in Scandinavia (as the Sun hides it in summer there). In Egyptian mythology, it was considered to be a pool of cow's milk associated with the fertility cow-goddess, Bat, while, among the many native peoples of North America, it bears different names such as The Scattered Stars for the Apache; Dog Tail for Cherokees; Trail of Bubbles for the Dakota people; Raven's Snowshoe Tracks for Inuits; and Flour and Ash for the Pima tribe. The Aborigines saw an Emu in the Sky, whose shape is delineated by light and dark regions of the Milky Way, and envisaged the small bright patches offset from Southern Milky Way as campsite lights of sky people. When first seen by European explorers, these bright offsets were named the Magellanic Clouds and are now known to be small galaxies orbiting ours.

Figure 4.1.1 Milky Way from two nights in Namibia using portable home gear
(Mark Hill, Neilston www.flickr.com/photos/astroscot2)

Figure 4.1.2 Andromeda Galaxy M 31 in visible light
(Graham Malcolm, Kilbarchan)

The Milky Way is now known to be the accumulated light from a vast number of stars (over 10 billion) and associated gas and dust, now designated as a galaxy. It spreads over distances of around 100,000 LY or about 25,000 times the distance to the nearest star – see below and Section 1.3. It is one of around an estimated 10^{11} galaxies scattered across the cosmos ranging in size from dwarfs, like the Magellanic Clouds, to giants, like our Milky Way and its near neighbour the Great Andromeda Galaxy. The latter, at 2.5 million LY, is the furthest object visible (in good dark skies) to the naked eye. The history of how the distances and nature of galaxies were first recognised is important as a key step toward today's view of the large-scale structure of oor (very!) big braw cosmos.

Invention of the telescope and its application to astronomy at once showed that the 'wandering' planets differ from the fixed stars in appearing not as points but as sharp disks of noticeable angular size. Later, it discovered various fixed objects of appreciable angular size but more or less fuzzy edged (nebulous) appearance. Some of these proved to be clusters of stars (see Section 3.3), so close together as to be unresolved in small telescopes while others, termed nebulae, are genuinely diffuse. Many of the latter are closely related to stellar life cycles (Section 3.4). The first systematic cataloguing and study of nebulae and clusters was by Charles Messier (France, 1730–1817), who assigned them Messier M Numbers still used today, like M42 for the Great Nebula in

Figure 4.1.3 Andromeda Galaxy M31 in UV light, dominated by energetic light from hot, young, massive stars unlike the visible light view in Figure 4.1.2
(NASA GALEX, JPL-Caltec apod.nasa.gov/apod/ap150724.html)

Orion's sword, M45 for the Pleiades Star Cluster in Taurus and M31 for the Andromeda Galaxy.

It was not until 11 February 1845 that the discovery was made by William Parsons, 3rd Earl of Rosse, Ireland, that there is a distinct class of diffuse nebulae which exhibit spiral shapes. This ultimately world-changing discovery was made within months of Rosse's first trial of his great 6-foot (1.8 m) telescope, the Leviathan of Parsonstown, the world's largest telescope from 1845–1917 (Figure 4.1.4). Being pre-photography, these observations were made by eye and recorded as drawings, including large numbers of individual stars resolved by the giant optics (Figure 4.1.5). Although today's technology enables vastly better images to be recorded digitally using much smaller telescopes that are affordable by amateurs (eg Figure 4.1.6), these drawings were exquisite and pioneering at that time. Among many other objects sketched in detail by Rosse was Messier M1, which he named the widely known Crab Nebula.

The discovery of large numbers of resolved stars in spiral nebulae generated a long debate as to whether spiral 'nebulae' are separate galaxies of stars like the Milky Way lying well outside it (Island Universes), or just one class of nebula within it. The only sure way to resolve this was to measure their distances. Even with today's best instrumentation, the parallax/triangulation method (see Chapter 1) can only extend to about 10 per cent of the Milky Way and, for greater distances we need to use a bright Standard Candle. The basic idea is that, if one sees how bright something appears and somehow knows how bright it really is, one can figure out how far away it is. In the case of astronomy, studies of nearby stars, whose distances are known by triangulation, allows inference of their absolute brightnesses (ie luminosities in watts). For example, Proxima Centauri, the nearest star, is found from its parallax to be about 276,000

Figure 4.1.4 The 6-foot Leviathan
Telescope of William Parsons,
3rd Earl of Rosse
(commons.wikimedia.org/wiki/File:Leviathan_of_
Parsonstown..._(8392017304).jpg)

times further from Earth than the Sun and it appears about 48 billion times fainter. Allowing for the spreading of light with distance (inverse square law) this implies that, in absolute terms, it is about 50 per cent more luminous than the Sun.

Based on such figures for a large sample of nearby stars, it is found that stellar luminosities are related to some of their other properties which are distance-independent such as their colours. For example, Proxima Centauri is a little bluer, since hotter, as well as more luminous than the Sun. This means that if we measure the colour of a star, we know its luminosity, making it a Standard Candle. So long as the Standard Candle relationship, in this case of colour to luminosity, holds true for distant as well as for nearby stars, seeing the colour of a very distant star tells us its absolute

Figure 4.1.5 Whirlpool
Galaxy M51 sketch by the
Earl of Rosse
(commons.wikimedia.org/wiki/
File:Astronomy_for_the_use_of_schools_
and_academies_(1882)_(14764435545).jpg)

Figure 4.1.6 Photo of
Whirlpool Galaxy M51
(Graham Malcolm, Kilbarchan)

Figure 4.1.7 Portraits of Leavitt (left),
Slipher (centre) and Hubble (right).
Key figures (among others) in the
discovery of cosmic scale and expansion
(Wikimedia Commons)

brightness – hence, given its apparent brightness, its distance.

In practice, there are much more reliable Standard Candles than stellar colour, the first being discovered by Henrietta Leavitt (1868–1921) of Harvard College Observatory, and proving immensely important in subsequent studies of the scale of our cosmos. Her study was of Cepheid Variable Stars, whose brightnesses and colours pulsate over periods of days to months with mean luminosities of 100–10,000 times that of our Sun, so visible at large distances. By observing thousands of Cepheids in the Magellanic Clouds, all nearly the same distance from us, Leavitt discovered their pulsation periods are correlated closely with their luminosities making Cepheids excellent Standard Candles for measuring quite large cosmic distances across the Milky Way and beyond.

Soon after Leavitt's discovery, two other momentous astrophysical breakthroughs were made by galaxy spectroscopist, Vesto Slipher (US, 1875–1969). A spectrum line comprises light emitted (or absorbed) by an atom at a very sharply defined wavelength (colour). The wavelength and intensity of spectrum lines can be used to detect and measure the abundances of chemical elements – ie what remote cosmic objects are made of. Somewhat surprisingly we find the same elements across the entire Universe, though their abundances vary significantly with location depending on the stage that stellar nucleosynthesis (Section 2.4) has reached there. Spectrum lines also carry invaluable information on the speed of their source toward or away from the observer via the Doppler effect (see Section 1).

In 1914, Vesto Slipher discovered spectrum line wavelength variations across the images of flattened disk spiral galaxies – see below – blue-shifted on one side of the galactic centre and red-shifted on the other. This shows the disk to be rotating and the speed of rotation allows the mass of the galaxy to be estimated from the need to keep the galactic material in gravitational orbit. Secondly, in 1912, Slipher measured Doppler shifts of lines from whole galaxies and showed that all but the nearest of them are receding from us, the recession speeds being larger for more distant (fainter, smaller looking) galaxies. The implication that our Universe is expanding was thus well-known to Slipher and colleagues well before 1917. Twelve years later, with his colleague, Milton Humason, Edwin Hubble (US 1889–1953) published his 'Velocity Law' paper, which later made him famous worldwide as the generally but inaccurately perceived discoverer of cosmic expansion.

The modest Slipher was not even mentioned in the paper, let alone made a co-author, despite the fact that almost all the Doppler speed axis values used by Hubble were from Slipher's work. What Hubble *did* do was use the Cepheid method to quantify the galaxy distance values on the other axis and show that galaxy recession speeds increase directly proportionally to distance.

4.2 Galaxies and the Extragalactic Universe

Though Hubble was *not* the discoverer of cosmic expansion, he was heavily involved in quantifying it and contributing to the study of galaxies and the extragalactic Universe in general. His 1929 paper on Cepheid measurement of the distance to Andromeda M31 is widely cited as the first firm proof of its lying far outside the Milky Way and being a galaxy in its own right. However, early in the 1920s, Lloyd Shapley and Ernst Öpik were already offering M31 distance values not unlike Hubble's 1929 'discovery' result. Hubble himself said as much, so perhaps the historical adulation of Hubble is as much due to our cult penchant for celebrity heroes as it is to personality trait differences between Slipher, Hubble and others. In any case, the hard facts are that in the 100 years following 1820 or so, the measured size of our world increased: (i) by a factor of around 10,000 from the Solar System to the nearest star – via the triangulation work of Thomas Henderson and others; (ii) by a roughly similar further factor using Leavitt's Cepheid method to reach across the Milky Way; (iii) by a further factor of around 20 to Andromeda, a total scaling up by a factor of around 2 billion. This already mind-blowing result was, however, to pale into insignifance as new brighter Standard Candle methods were discovered, such as the peak brightness of SN explosions.

From these, it was eventually realised that galaxies after galaxies, fainter and fainter, receding faster and faster, are spread out, though clustering on large scales (Figure 4.2.1), to the edge of the visible Universe at around 14 billion LY. There, the Doppler reddening makes them too faint to be seen since they are expanding away from us at the speed of light. If you took a week to read this book, during that time, a further 200 billion km worth of the Universe would have accelerated out of our visible Universe. The total volume contains around 100 billion galaxies, with up to 100 billion stars in each, a stellar total 10^{22} stars. This can be visualised somewhat by looking at the mosaic of very long exposure Hubble Telescope Deep Field (HDF) images in Figure 4.2.2, which covers only about one 24 millionth of the sky (about the area hidden by one grain of sand held at

Figure 4.2.1 Virgo cluster of galaxies (above)
(Rogelio Bernal Andreo, www.DeepSkyColors.com apod.nasa.gov/apod/ap150804.html)
Figure 4.2.2 Hubble deep field
(R Williams STScI, NASA HST HDF Team, Wiki Commons: HubbleDeepField.800px.jpg)

arm's length). Only a small number of objects in this figure are foreground stars. The rest are all very distant galaxies at the farthest reaches of the cosmos.

As well as the large recession speeds caused by large-scale stretching of space by cosmic expansion, the Doppler effect reveals a number of other important motions within galaxies, the dominant motion depending on (or determining) galaxy type or shape. The commonest types (Hubble Classification) are Spiral and, rarer and mostly fainter, Elliptical Galaxies. Spirals are quite flattened disks, with the stars and gas moving in near circular orbits but showing spiral (Catherine-wheel) distribution of the material (some Barred Spirals have a central straight bar). Ellipticals are shaped roughly like rugby-balls with varying degrees of elongation, contain less gas, and have stellar orbits fairly random in speed and directions. They usually have older stars and less gas and dust than Spirals. The origins, evolution and relationship of the various kinds, with their differing structure and contents, is still not clearly understood. As well as the variety present in intrinsic galaxy structure, different angles of view introduce widely varying appearance to intrinsically identical galaxies, especially Spirals seen edge-on versus face-on (see Figure 4.2.3). Comparison of the last panel (vi) of this Figure, showing Centaurus-A as taken by Scottish amateur Mark Hill with the Hubble Telescope close-up in Figure 4.2.4, is a great example of what amateurs can achieve but brings out how much more large aperture telescopes can attain, especially when used to access wavelengths only accessible above the atmosphere.

Since similar types of galaxy are seen at many different distances, their structures must be long-lived. In particular, they must be massive enough for their gravity to prevent escape of their fast-moving stars and gas. As already mentioned, the rotational speeds of flat spiral galaxies had already been measured by Slipher very early in the 20th century. However, it was only around 1960 that Vera Rubin (US, 1928–2016), Kent Ford and others made the dramatic discovery that the mass in Spiral Galaxies needed to retain their stars in orbits is far higher (about five times) than the amount of mass we actually see as visible matter via its emission (stars) or absorption (cool gas and dust) of radiation. This discrepancy can only be explained by invoking the existence of invisible material – dark matter – or by conjecturing some deviation on large scales from Newton's law of gravitation. The need for dark matter had been proposed much earlier – around 1930 – by Fritz Zwicky (Switzerland, 1898–1974) in the context of clusters of galaxies, where he found that the mean random speed of galaxies in them is far too fast for the cluster to be held together by the gravity of the matter visible to us. Whether and how much dark matter is present in Elliptical Galaxies is unclear, as is the relationship of Ellipticals to Spirals and the origins of each. In fact, it is still even an open issue whether galaxies or stars form first. Does the self-gravitational collapse of huge clouds of gas into protogalaxies lead on to subsequent star formation by collapse of small clouds ('top down' scenario), or do stars form first and help galaxies to form around them ('bottom up' scenario)? At the time of writing, a claim has just been announced that very small but very old galaxies (over 10 billion years) have been observed near the much more recently formed Milky Way. Whether this will help clarify things or muddy the waters, only time will tell.

Figure 4.2.3 Diversity of galaxies: Top left to bottom right (*not* to scale) –
i) M33 The Triangulum spiral galaxy seen obliquely – cf. Andromeda spiral Figure 4.1.2 seen nearly edge-on
and the Milky Way spiral in Figures 4.0.1 and 4.1.1 seen edge-on from inside. M33 is the second nearest
galaxy (after M31) to the Milky Way and part of an interacting local group
(Graham Malcolm, Kilbarchan)

ii) M101 the Pinwheel spiral galaxy seen almost face on. Roughly ten times more distant than M31 and M33
(Graham Malcolm, Kilbarchan)

iii) Barred Spiral NGC 1300 seen almost face on.
(NASA, ESA and The Hubble Heritage Team STScI/AURA)/; Wikimedia Commons)

iv) Giant Elliptical Galaxy ESO 325-G004 with other smaller diffuse ellipticals across the field. Bright points
seen near it are mostly compact globular star clusters
(NASA, ESA and HST Heritage Team (STScI/AURA); J Blakeslee (Washington SU); Wikimedia Commons)

v) Dwarf Irregular Galaxy UGC_820 (ESA/HST & NASA
(www.spacetelescope.org/images/potw1510a/)

vi) Centaurus A (NGC 5128): See Figure 4.2.4 for close-up and details
(Mark Hill, Neilston)

Chicken and the Egg;
The Galaxy or the Starn?
Wha kens whit cam first...

An important element of galaxy research is to understand the similarities and differences between the different types and appearances of galaxies in terms of their size, composition and age. This involves studying and categorising large data sets of galaxy images, which has been greatly aided by the Citizen Science Galaxy Zoo project (zoo1.galaxyzoo.org) of the BBC's *Sky at Night* programme and led by Professor Chris Lintott of Oxford. At one time, it was thought that Ellipticals and Spiral structured disk galaxies were an evolutionary sequence. However, both contain a comparable fraction of old stars, so should be of similar age, though Ellipticals do have lower abundances of heavy elements which could have indicated an age difference. One theory about this is that Ellipticals form from giant gas clouds much more tenuous and lacking in gas than Spirals. As a gas cloud collapses, most of the friction is due to the extended dilute gas rather than collisions between the widely separated stars. Thus, a

Figure 4.2.4 Centaurus A (left) closest ever look into the dust of the Cen A galaxy by HST WFC3, combining features in the visible, with UV from young stars and dust penetrating near-IR light
(NASA/ESA Hubble Heritage, STScI/AURA -ESA/Hubble Collaboration. R O'Connell (U. Virginia) and the WFC3 Scientific Oversight Committee)

Figure 4.2.5 *Birth of a Galaxy* – oil on canvas board 10x8 inches
(JC Brown 1975)

Figure 4.2.6 Mice (NGC 4676) galaxies (above)
(NASA, H Ford (JHU), G Illingworth (UCSC/LO), M Clampin (STScI), G Hartig (STScI), the ACS Science Team, ESA commons.wikimedia.org/w/index. php?curid=539276i/Interacting_galaxy#/media/File:NGC4676.jpg)

Figure 4.2.7 Galactic gravitational dance (right)
(NASA JPL/Caltech (www.galex.caltech.edu/media/glx2006-03r_img03.html)

collapsing tenuous low-mass gas cloud will tend to form stars but will not have enough gas friction to flatten the rotating gas cloud into a disk, nor will it leave enough gas to create a second generation of stars. More massive denser initial gas clouds would exert far more friction thus concentrating the high angular momentum into a disk while leaving lots of gas to form second-generation stars with high abundance of heavy elements. Another suggestion is that low angular momentum Ellipticals might be the products of collision/merger of flat Spirals, each with high angular momentum but with different spin directions. In such encounters, there could be much cancellation of their directed angular momenta, leaving an Elliptical structure.

In any case, it seems that gravitational interaction of galaxies with their nearest neighbours is important, despite the great separation between their individual stars (compared with their sizes). The systematic Hubble recession of galaxies only holds over large distances. Random motions of nearby galaxies can make them approach each other, as found by Slipher in 1913 for our nearest neighbour, M31 in Andromeda. It is approaching the Milky Way at around 100 km/sec and they will collide in about 5 billion years, as has happened many times for other galaxies in the past. In such collisions, there are very few real 'hard' impacts of stars with stars – they are so very far apart – but a lot of gas friction slows things and gravitational interaction of stars ejects some out of the system.

Many such galaxy collisions and their products are seen across the cosmos (Figures 4.2.6 and 4.2.7). The Mice Galaxies in 4.2.6 are an iconic image. In 4.2.7, NASA's Galaxy Evolution Explorer, GALEX, shows in UV light spiral NGC 1512 and Elliptical NGC 1510 just 68,000 LY apart (less than the size of the Milky Way) so in a close gravitational encounter. Two tightly wound spiral arm segments make the blue inner ring of NGC 1512, while its outer spiral arm is being gravitationally distorted by NGC 1510. On smaller scales, there are also quite frequent interactions of small galaxies and larger galaxies with ones into which the latter are assimilated, a process termed galactic cannibalism!

In my painting, *Birth of a Galaxy*, in Figure 4.2.5 above, I meant to convey the idea of the birth of structures like galaxies in the Universe. My choice of the Whirlpool Galaxy M51 was partly based on its somewhat sperm or serpent-like shape, though the modern view is that the 'head' is, in fact, a small galaxy being cannibalised by the large one!

To close, one should note that there is a creative side to the processes of galaxy collision and cannibalism, although they look pretty violent – especially in the slightly blood and guts look of Figure 4.2.8, which shows the colliding Antenna Galaxies. But the fact here is that, while two galaxies are reduced to one with ejection of many stars and much gas, the process of gas compression involved also enhances the gas cloud collapse and star creation rate throughout much of the region, thus giving birth as well as death.

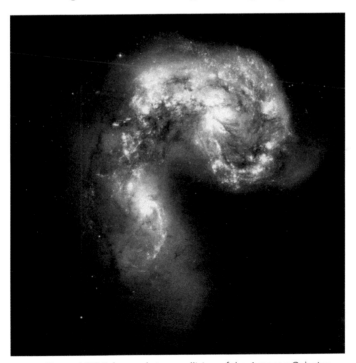

Figure 4.2.8 Advanced stage collision of the Antenna Galaxies
(NASA, ESA, Hubble Heritage Team (STScI /AURA)-ESA/Hubble Collaboraton www.nasa.gov/multimedia/imagegallery/image_feature_1086.html)

Galaxies that merge;
'Galactic Cannibalism'!
Wha wid hae thunk it?!

4.3 Cosmic Expansion, Dark Matter, Dark Energy and the Multiverse

We have seen that the rotational speed of Spiral galaxy disks and the random speeds of galaxies in galaxy clusters are both much too high to be contained by the normal gravity of all their visible matter. The survival of these structures demands either an increase of gravity over Newton's inverse square law at large distances or the presence of invisible dark matter. The latter is the orthodox view of scientists, although decades of expensive efforts to detect dark matter directly on Earth (for example, down Yorkshire's Boulby mine) have thus far failed.

On much larger scales, accurate Doppler measurements of the cosmic expansion speed, $v(r)$, as a function of distance, using very bright SN as Standard Candles to measure distances, r, can tell us about the forces acting over cosmological distances. For instance, at large distances, $v(r)$ would be expected to fall below the Hubble straight line proportionality law if the only force acting were the gravity of the cosmos retarding the expansion. The decrease in acceleration would indicate the mass of the Universe and whether or not it would be sufficient eventually to stop the receding matter and eventually make it fall back (see Chapter 8). That levelling out of $v(r)$ can be seen after the inflation phase in the middle of the schematic Big Bang timeline of Figure 1.3.1 in Chapter 1. This question was central to cosmology in the late 1900s.

Figure 4.3.1 Gravitational lensing – an exaggerated simulation
(NASA, ACS Team, Richard Bouwens (UCO/Lick Obs.
www.nasa.gov/multimedia/imagegallery/image_feature_968.html)

Figure 4.3.2 Cosmic microwave background
(NASA WMAP Science Team map.gsfc.nasa.gov/media/121238/index.html)

However, in the 1990s, precision SN studies yielded the shock result that the increase of v with r actually accelerates, leading to award of the 2011 Nobel Prize in Physics to Saul Perlmutter, Brian Schmidt and Adam Riess. This late-stage accelerating Universe can be seen toward the right end of the schematic Big Bang timeline of Figure 1.3.1 as the belling outward at late times. Two decades on, there is still debate concerning the result and its interpretation in terms of the presence on large scales of a hitherto unknown outward force (like a compressed spring or sponge) embedded in space. The source of this force has been called dark energy though, like dark matter, there is no direct (eg laboratory) evidence for its existence and ideas as to its nature are even more scant than for dark matter. Being at present pretty much ad hoc conjectures, these concepts, like mediaeval epicyclic orbits, are not very amenable to making testable predictions. Thus, for now at least, they have to be thought of as near the margins of science and science fiction, alongside fanciful concepts like our Universe being one of an infinite number making up a multiverse endless in space and time.

Another fairly recent discovery about galaxies is that most Spiral galaxies and some Ellipticals harbour supermassive black holes in their centres, sometimes associated with activity of various kinds including strong radio emission, traditionally called Active Galactic Nuclei (AGNs) or Quasars. Such concentrations of mass and their strong gravity contribute to the now widely seen phenomenon of gravitational bending and lensing predicted by Einstein. The gravitational lensing phenomenon is where the light bending by the gravity of a high gravity object forms distorted multiple images of a distant object behind it. An exaggerated but illustrative version of this is shown by the computer simulation results in Figure 4.3.1, where the view of a distant cluster of galaxies has been gravitationally lensed by a foreground cluster of galaxies centred on a large Elliptical galaxy. Galaxies far behind the cluster are lensed into the many concentric arcs by the large gravity of dark matter in the cluster.

The effect was first seen in reality by Arthur Eddington as displacement by the Sun's gravity of stars near the Sun in the sky during the 1919 solar eclipse test of General Relativity. The process of gravitational lensing was already mentioned in Section 2.6 in connection with exoplanet detection, but there the lensing angles were very small and the lensing only seen via its stellar brightness amplifying effect, not imaging. This is called gravitational microlensing.

Finally, to recap what was said in Chapter 1, a crucial item in many recent studies of large-scale structure of the cosmos is the CMBR, predicted by Alpherin, Herman and Gamow in 1948, and detected by Penzias and Wilson in 1964, earning them a Nobel Prize in 1978. This is remnant radiation from the Big Bang, cooled by cosmic expansion to 2.7 K leaving tiny ripples in temperature across the whole sky. The WMAP data in Figure 4.3.2 show the amplitude and scale of today's CMBR temperature ripples (as colour differences) across the sky to be just 200 microkelvin (around one five thousandth of a degree Celsius) deviations from the mean. These are believed to arise from tiny spatial irregularities in primordial cosmic matter originating 13.77 billion years ago in Big Bang fluctuations that condensed gravitationally into the large-scale cosmic structure like the galaxies we now see. Studying those ripples is thus crucial in testing cosmological theories.

The problem is that the CMBR only comes from the furthest cosmic distance (earliest time) to which we can see in electromagnetic radiation. The matter beyond that is so dense that it is opaque to light and to all parts of the electromagnetic spectrum (see Section 1.3). However, after half a century of effort, the birth in 2015 of the new discipline of gravitational wave astronomy may ultimately let us probe behind that electromagnetic curtain.

Gravitational waves, predicted by Einstein, are travelling ripples in space time and are emitted by accelerating masses. They are somewhat analogous to the electromagnetic waves predicted by James Clerk Maxwell (Section 5.8), which are emitted by accelerated charges. However, gravitational waves interact much more weakly with matter. This makes them hard to detect but allows us to see inside electromagnetically opaque things like stars. Only recently have large hypersensitive instruments managed to detect them. One such of several worldwide is the US-based LIGO. LIGO comprises a pair of detectors on opposite sides of the country, each with two perpendicular 4 km long vacuum tubes carrying intense laser beams back and forth between ultra-reflective mirrors to measure the shift of test masses induced by the waves. It is able to measure movements 1/10,000 the size of an atomic nucleus such as arrive as distortions of space itself by the gravitational waves from extremely violent and remote cosmic gravitational events. The total power in the LIGO laser beams used to measure these tiny spatial distortions is a mere 200 W. (Previously, gravitational wave emission had only been inferred indirectly, via their effect on the timing of pulsars in binary star systems.)

CHAPTER 5
Some Early Great Scots Astronomers

Wha's like us?! Damn't few!
These are deid, an yet they leeve,
In planets, starns, munes...

5.0 Today's Scottish Astronomy Scene

DESPITE OUR OVERLY maligned but nonetheless rather cloudy weather, the people of Scotland, and even more so diaspora Scots across the globe, have long been fascinated by the heavens and were active in the study of them (see Section 6.1 for the prehistory). A little known fact is that many of the principles of spaceflight and rocket motion in a vacuum had been laid down by William Leitch (1814–64) of Rothesay, Isle of Bute (www.wikipedia.org/wiki/William_Leitch_(scientist)), long before the usually cited greats – Konstantin Tsiolkovsky (1903), Robert H Goddard (1914) and Hermann Oberth (1923). Turning to recent times, while no great fans of league tables, we cannot resist some boasts such as the following: based on place of birth, Scotland has more Nobel Laureates per head of population than any other country, except Sweden and Switzerland; a recent THES survey, which ranked nations by auditable quality publications in astronomy and space research, placed Scotland, Israel and Canada at the top of the table, per capita, with seven small countries in the top ten.

Some major recent astronomical highlights in Scotland include: the establishment of the highly successful UK Centre for Astrobiology in

Figure 5.0.1 The Sky is not the Limit, 48x24 inches oil on canvas, Dugald Cameron 2012
(Dugald Cameron and the Scottish Space School)

Edinburgh; Peter Higgs' Nobel Prize; the construction of the key MIRI instrument for NASA's James Webb Space Telescope (HST successor) by a European consortium, led by the Royal Observatory Edinburgh; Glasgow's Clyde Space company has the world's highest rate of production (three per month) of general use micro-satellites (Cubesats); the announcement in July 2018 by the UK Space Agency that the UK's first vertical launch spaceport site will be at A'Mhoine, Sutherland; the deep involvement of the University of Glasgow's Institute of Gravitational Research in the American LIGO and other gravitational wave detection instruments.

On the public and amateur fronts, Scotland has four public observatories, numerous astronomical societies and many extremely able astrophotographers, several of whom have kindly contributed superb images to this book, almost all of them taken from Scottish sites. Scotland now has many dark sky observing locations recognised by the International Dark-Sky Association (IDA). Coll has Dark Sky Island status, while the Galloway Forest and Tomintoul and Glenlivet Cairngorm Dark Sky Parks have IDA Gold status. The former opened in 2012 and contains near its boundary the Scottish Dark Sky (public) Observatory – see Section 7.4 – and the latter, which opened in November 2018, is the world's most northerly of such parks. Today's fine state of Scottish astronomy rests partly on the shoulders of the many great Scottish astronomers of bygone centuries, of whom the following Sections 5.1–5.9 give a sample.

A number of the above highlights and other Scottish space success stories mentioned elsewhere in this volume were also contained in the fine 2012 painting *The Sky is Not the Limit* (Figure 5.0.1) by Professor Dugald Cameron OBE, FCSD, FRSA, Director of Glasgow School of Art 1991–99. This work celebrated the highly successful Scottish Space School series that was conceived and run by the University of Strathclyde. His own summary of its contents and concepts in 2012 reads as follows:

> On a background of the Universe, the Space Shuttle OV-105 *Endeavour*, which, on mission STS 130 in February 2010, took a miniature edition of the poetry of Robert Burns into space. Representative developments in aerospace are shown – the solar sails of Professor Colin McInnes and the UKube satellite of Clyde Space.
>
> As a new dawn rises over the West of Scotland, a representation of those young people who have enjoyed the Scottish Space School over its ten and more years of existence is depicted along with the coat of arms of the University of Strathclyde and a relevant equation of motion, devised in the university. Percy Pilcher's glider, is shown in which he made the first repeated heavier than air flights in the United Kingdom at Cardross in 1895 as is the 'Sputnik', thistle symbol of the xx Planetary conference held at Strathclyde University in 2007.

In the background against a myriad of galaxies and constellations is the great Orion, 'The Hunter' bestriding the imagery. The light we see from its M42/43 nebulae in 2012, began its journey through space around the time when Kentigern was camped by the Molendinar burn from where arose, the City of Glasgow.

Science, Technology, Engineering and Mathematics are incorporated in the image and, of course the Saltire, which King Angus adopted from seeing clouds making a cross against an azure sky at the Battle of Athelstaneford in AD832.

5.1 James Gregory (1638–75)

James Gregory was a great Scottish mathematician and astronomer, sometimes known as the Scottish Newton. Among his many accomplishments were helping to lay the mathematical foundations of calculus and inventing the first (Gregorian) possible layout for a mirror-based telescope. These avoid the chromatic (colour) aberration that afflicts lens telescopes, since the colours of light disperse as it passes through glass. His birth preceded Sir Isaac Newton's by four years and his reflecting telescope design in 1663 preceded the Newtonian one. However, the latter was easier to construct and was realised by Newton in 1668, five years before Robert Hooke built a real Gregorian. He also invented dispersion of light by diffraction gratings based on seeing the effect of diffraction through the strands of bird feathers. In 1668, Gregory became Regius Professor of Mathematics at the University of St Andrews, aged 30 but only survived a further 7 years, a lifespan well under half of Newton's.

Many Gregorian telescopes were later produced by the famous Scottish telescope maker (and mathematician), James Short (1710–68, born in Aberdeenshire). During his 35-year career in Edinburgh and London, he created 1360 telescopes, the style of one of which features in the Astronomer Royal for Scotland Coat of Arms, devised in 1996 (see Section 7.5).

5.2 Alexander Wilson (1714–86)

A polymath and colossus of creativity, Alexander Wilson was born in St Andrews. After graduation from the University of St Andrews in 1773, Wilson was apprenticed to a local surgeon and learned the arts of glass-blowing. After an unproductive business sojourn in London, he returned to the Glasgow area of Scotland, successfully established businesses in the type-founding trade and in glass-blown instrument production. These included accurate thermometers and specific gravity beads for measuring

Figure 5.2.1 Alexander Wilson, 1st Regius Chair of Astronomy, University of Glasgow
(James Babington Smith, Wilson descendant)

the proof of spirits. Through business connections in the University, he provided the famous Greek fonts (Fontana, Scotch Roman and Wilson Greek) for the exquisite production by the Foulis press of Homer's *Iliad* and *Odyssey*. He participated in philosophical discussions at his home and in the City and, with college friend Thomas Melville, conducted experiments with kite-borne thermometer arrays to measure atmospheric temperature structure, predating the kite physics experiments allegedly carried out by Benjamin Franklin.

In 1756, Alexander Macfarlane of Jamaica bequeathed his astronomical instruments to his Alma Mater and the Old Glasgow College opened the Macfarlane Observatory in 1757, east of the High Street. The instruments were refurbished by a College instrument maker who was none other than the now famous James Watt (1736–1819) whose studies of energy and of power (energy/sec) eventually led to the modern unit of power – the watt (w) – being named after him. The Duke of Argyll, having been very impressed by Wilson's excellent instrumentation skills and therefore anxious to keep them and his business acumen in Glasgow, initiated an appointment in 1760 for Wilson to run the observatory as First Regius Professor of Practical Astronomy. On appointment, Wilson moved his type-founding business to premises in the College.

In 1769, Wilson made his most significant contribution to astronomy. While observing a Sunspot traversing the rotating solar disk, he realised that, when seen close to the solar limb, the spot appeared to involve a 'depression' below the Sun's regular surface, the dark central spot umbra progressively disappearing because it was becoming hidden by a higher rim around it. This is something one can explain to lay audiences today but, in Wilson's time, the idea of such structure on the Sun was very radical. The Royal Society of London published his geometrically convincing drawings and his more speculative interpretation that the Sun's 'surface' has vertical structure, with a hotter and more luminous outer layer covering a darker globe. This interpretation was accepted for over a century. Though his theory about the darker globe below was flawed, the Wilson Effect is still discussed today and his paper can be seen as the birth of modern solar atmospheric physics. This work won an essay prize and gold medal from the Royal Society of Copenhagen and led to Wilson's making Sun-mimicking apparatus in the observatory grounds. At the Macfarlane Observatory, Wilson also observed the transits of Venus across the solar disk in 1761 and 1769, and measured positions and eclipse times associated with Jupiter's satellites.

Figure 5.2.2 Wilson Effect drawing, 1769
(University of Glasgow Astronomy archives per JC Brown)

In 1777, Wilson wrote an important tract on the Universality of Gravity affecting the motion of stars. This promoted the idea that Newton's Laws are applicable beyond the Solar System and that the stars do not collapse on each other under gravity because they are in motion. This notion was also being considered by Sir William Herschel. When Herschel published his famous work on this a few years later, he graciously acknowledged Wilson's contribution. Wilson's thermometers were later loaned by his son Patrick to Herschel and became instrumental in Herschel's 1800 discovery of infra-red radiation.

In 2017, Wilson's many contributions to physics were recognised in the form of an Institute of Physics Blue Plaque installation at the University of Glasgow's Acre Road teaching observatory (Figure 5.2.3). A very interesting early biography of Wilson is to be found at www.biodiversitylibrary.org/item/20116#page/1/mode/1up.

Sadly, the site of Wilson's original observatory has long since vanished, along with almost every other trace there of the Old College of Glasgow, with the Old College Bar in High Street (built in 1515) threatened by the wrecking ball of developers. On today's University Gilmorehill site (Gilbert Scott Building, 1870), Pearce Lodge and the Lion and Unicorn Staircase were transferred from the High Street site. The University of Glasgow archives and the National Museum of Scotland both still preserve some Wilson instrumentation artefacts, from telescopes to thermometers and SG beads. The only remaining artefacts of the University of Glasgow Observatory buildings from the Old College era seem to be a plaque displaying a star within a triangle within an Ouroboros (tail-eating serpent), presumably symbolising astronomy, geometry and the endless circle of knowledge. This originally adorned the Glasgow Horselethill Observatory in the 1800s. A painted version of it was installed in the current Acre Road teaching observatory when it was opened in 1969 – it is just obliquely visible to the right of the doorway in Figure 5.2.3. A stylised version of that became the symbol of the Astronomy & Astrophysics (A&A) Research Group after 1986, when the Departments of Astronomy and of Natural Philosophy amalgamated to become what is now the School of Physics and Astronomy. In 1989, the long lost original stone Ouroboros plaque was found by chance, lying moss-covered in a campus garden and is now resplendent by the current main entrance to the Kelvin Building. Figure 5.2.4 shows this alongside the A&A logo based on the painted Acre Road version.

Figure 5.2.3 Wilson Blue Plaque (IoP Scotland) unveiling on 28 March 2017 at University Acre Road Observatory site. Left to right: Rab Wilson (Scots Scriever/Writer in Robert Burns Birthplace Museum), Professor JC Brown (Astronomer Royal for Scotland), Dr David Clarke (former Observatory Director), James Babington Smith (Wilson descendant) and Professor Sir Anton Muscatelli (University of Glasgow Principal)
(University of Glasgow Photo Unit)

Figure 5.2.4 University of Glasgow Astronomy Ouroboros emblem. Original stone version from Horselethill Observatory, now embedded in the wall of the Kelvin Building, home of the School of Physics and Astronomy, and logo of the present Astronomy and Astrophysics Research Group schematically based on the painted emblem at Acre Road Teaching Observatory
(JC Brown)

The Sairchin Mind

Alexander Wilson 1714–86,
1st Regius Professor of Astronomy, University of Glasgow

Wilson chairts the muivement o the heivins,
Genius aye maun dae whit it must,
This polymath unpicks the Universe,
Wi practised haund, a gleg an eident ee.

A lad o pairts wi naitrel ingyne;
Ye craft in gless, in bress, in wuid…
Graduating scales, tae measuir whit?
That which until then hud no bin measuirt.

Thermometers attached tae paper kites,
Sail ayont the cloods tae dizzyin heichts,
While weans luik oan, bumbazed at sic ongauns,
Ye brack the spell o superstitious haivers.

Whit hauds the starns frae fawin frae the sky?
Ye ettle aathing spins around its centre,
An fir yer time ye're no faur aff the mark;
Yet ne'er cuid reck the holes in nichtime's sark.

Thochts that until then hud no bin thocht;
Archimedan rays that brunt a fleet,
Whiles in a nod tae ither classic lore,
Ye e'en set type tae publish Homer's lays;

Fontana, Scotch Roman, and Wilson Greek,
Yer nem leeves oan in uncial curlicues,
Whase sweep an bend wimple syne then jouk,
Tae cairt minds back tae fabled Ithaca.

Wi Watt an Black ye set tae buildin clocks,
Mair accurate than ony in their day,
Construct an astronomical telescope,
Tae spier anent the truth o Newton's laws,

Syne map the mysteries o thon gowden ba,
That gies life tae aathing upo the yird,
Markin fremmit motes oan Phoebus' face,
Cannily recordin in yer notebuik.

In manner frienly, gentle aye an kind,
A mind unsated, curious tae the last,
Roused bi fiddle, or Music o the Spheres,
'Brisk and lively as ever to dance a reel!'

Figure 5.3.1 Sir Thomas Makdougal Brisbane, Portrait by John Watson Gordon
(Royal Society of Edinburgh)

5.3 Sir Thomas Makdougall Brisbane (1773–1860)

Thomas Makdougall Brisbane, born near Largs, was educated at the University of Edinburgh in astronomy and mathematics and, in his initial military career, won many distinctions in action across the world, rising to the rank of Major General (1813). From 1821–25, he was appointed Governor of New South Wales, a role in which he was less successful in part because of ongoing infighting among rivals and factions, but he made many useful improvements in the area and helped establish a new penal colony at a Moreton Bay site near what became the city of Brisbane, named after him.

Thomas retained a keen interest in astronomy throughout his life, establishing in 1808 an observatory at Brisbane House near Largs including the Three Sisters – massive pillars still to be found in downtown Largs, which bore small lights used for accurate alignment of the telescopes up the hillside. From this observatory, he contributed much to advances in time-keeping, navigation and accurate stellar mapping. On arrival in NSW with his equipment, he and his two assistants, Carl Rümker and James Dunlop, established the first properly equipped Australian observatory at Parammatta, creating an atlas of 7,385 Southern Hemisphere stars, rediscovering Encke's comet in 1822 and setting up the world's first geomagnetic observatory.

Thomas returned to Largs in 1826, was eulogised in 1828 by Sir John Herschel when presenting him with the RAS Gold Medal and was elected Fellow of the Royal Societies of London and of Edinburgh. He was President of the latter for no less than 28 years! In 1836, he was made a Baronet and offered command of British troops in Canada, then the chief command of those in India in 1838. He declined both and continued his valuable astronomical researches. His activities included setting up optical and magnetic observatories at Makerstoun, acting as a champion for science in general and being instrumental in the creation of the post of Astronomer Royal for Scotland and appointing its first incumbent, Thomas Henderson (see Section 5.5). Currently Brisbane Observatory Trust is in the process of clearing, consolidating and preserving the ruins of Brisbane Largs Observatory, shown prior to and after their sterling work in Figure 5.3.2.

Figure 5.3.2 Brisbane Largs Observatory – remains prior to and post-clearance work by Brisbane Observatory Trust.
(JC Brown and Dr MI Brown)

5.4 Mary Somerville (1780–1872)

Figure 5.4.1 Mary Somerville (left) John Jackson portrait; (right) Self-portrait
(Principal and Fellows of Somerville College, Oxford)

Mary Somerville was born Mary Fairfax in Jedburgh, fifth of seven children, of whom four survived beyond early childhood, her two brothers receiving a good education while Mary's little early education was in reading, but not writing, from her mother. As a young child, she 'spent the clear, cold nights at her window, watching the starlit heavens'. Finally, at age ten, she was sent to boarding school, but was mostly self-educated by reading every book in sight after she escaped from there 'like a wild animal escaped out of a cage', as she put it.

Fortunately for Mary, unlike most of her family who wanted her to learn needlework, an uncle encouraged her to learn widely. When around 13 years old, she by chance became interested in mathematics via Alexander Nasmyth who gave her painting lessons and talked of perspective and Euclidean geometry. This intense interest received no support from either her family nor her allegedly misogynistic naval husband and, only after his early death and her return to Scotland, was it allowed to flourish in a new circle of friends including Professor John Playfair (natural philosophy) and in correspondence with William Wallace (mathematics). Some of that correspondence addressed the Mathematical Repository of deep maths problems and, by 1811, Mary had won a silver medal for solving one of these to do with time. By then, she was reading Newton's *Principia*, Laplace's *Mécanique Céleste* and other texts in mathematical astronomy. She shared a passion for science with her second husband, William Somerville, eventual FRS, and they mingled with physicists like Playfair and Brewster in Edinburgh, with William, Caroline and John Herschel in London, and many European visitors.

Launched into this new world, Mary published her first research paper, regarding the solar spectrum, in 1826. In 1827, she was asked by the Royal Society to translate Laplace's *Mécanique Céleste* and, in doing so, she explained much of the content which was previously unfamiliar to most British mathematicians. The resulting work, *The Mechanism of the Heavens*, appeared in 1831 and was a resounding success, although she had not entirely escaped gender prejudice. A later Principal of St Andrews recorded in his notebook that 'it requires a moment's reflection to be aware that one is hearing something very extraordinary from the mouth of a woman.' Despite such male chauvinism, Mary's next book, *On the Connexion of the Physical Sciences* (1834), was also very well received, running to many editions, the 6th being where she postulated that a hypothetical planet was perturbing Uranus. This led John Adams to his subsequent discovery of Neptune (in parallel with Urbain Le Verrier).

Mary Somerville's works are known to have influenced James Clerk Maxwell and she was described by him as 'certainly the most extraordinary woman in Europe – a mathematician of the very first rank', while of herself later in life she said, 'Sometimes I find [mathematical problems] difficult, but my old obstinacy remains, for if I do not succeed today, I attack them again on the morrow.'

Her memory has been immortalised throughout history in the naming of main-belt asteroid, 5771 Somerville (1987 STI); the Somerville lunar crater in the eastern part of the Moon; Somerville College, Oxford; and Somerville House in Burntisland, Fife. Mary Somerville's achievements were recognised recently in the UK by her portrait adorning the new Polymer Royal Bank of Scotland ten pound note (www.rbs.com/heritage/subjects/our-banknotes/current-issue-notes.html).

Mary Somerville

*'In astronomy, we perceive the operation of a force which is mixed up with
everything that exists in the heavens or on Earth; which pervades every atom,
rules the motions of animate and inanimate beings and is as sensible in the
descent of a rain-drop as in the falls of Niagara; in the weight of the air, as in the
periods of the Moon.'*—Mary Somerville, Scots Astronomer (1780–1872)

Somerville College, Oxford, nemmed fir Mary.
A crater oan the Mune, nemmed fir Mary.
An asteroid belt, nemmed fir Mary.
JMW Turner, dumfounert bi Mary.

In an age whan lassies held their wheesht,
An lairnin wis a male presairve,
You gaithert fir yersel a hairst o lear,
Syne glaumed at aa the starns in heivin.

The statues, the medals, thae biggins nemmed,
The great an guid an company kept;
James Clerk Maxwell, Wattie Scott,
Babbage, Milbanke, an Ada Lovelace...
Eminent Victorians aa!

Desairvedly ye kythe in ilka poacket,
The Rose o Jedburgh's speirin smile,
A gaze o slee inspection;
Whit question's formin oan yer lips?

Whiles nou ye're whaur ye ettled aye tae be,
An aabody kens yer warth;
Twenty-fowr mile abune the bonnie yirth,
Newhill an St Stephen's weans hae launched ye,
Tae float upon the mantle o the cosmos.

Short an lang ye hang, suspended,
Weichtless as air,
The yae ee oan oor vast blue orb,
The tither ettles infinity,
Syne dour auld gravity wins ye back,
An ye faa lik a drap o rain.

A Wee Message Frae Mary Somerville

Micht no a lassie glaum at aa the starns,
An hae a *richt* tae write as weel as read?
Rax oot an tak whit lads hae aye bin gied,
Years syne they'll rue lear tint frae lasses harns!
Sae thon's the task that ah masel did ettle;
Obleegin uncle Tammas tae his credit,
Must hae seen some spark that gien me merit,
Encouragin, emboldenin ma mettle.
Reminiscin nou here, ayont time,
Vyin wi the Universe itsel,
Ilk microscopic molecule compels
Lagamachies an theories aa tae chime!
Lassies nou ye've aa ah wrocht tae win;
Expand yer minds, an nevir aince gie in.

5.5 Thomas Henderson (1798–1844)

Born in Dundee, Thomas Henderson excelled in all his school subjects and received special tuition in mathematics, physics and chemistry by the gifted rector of the High School, Thomas Duncan, later professor of mathematics at the University of St Andrews. At the age of 15, he became a legal secretary, first in Dundee, then Edinburgh, and finally secretary to the Lord Advocate. With a keen (self-taught) interest in astronomy, he joined the Edinburgh Astronomical Institution and, when south on legal business, he met eminent astronomers John Herschel and George Airy, and was given full access to Sir James South's magnificent observatory at Camden. At that time, he invented a new method of determining longitude from lunar occultation timings.

Based on these credentials, Henderson was appointed as His Majesty's Astronomer at the Cape backed by Airy and Sir Thomas Makdougall Brisbane (see Section 5.3). Despite problems with colleagues and equipment there, he returned to Edinburgh with a large successful collection of data – a catalogue of 172 southern stars, data on Encke's and Biela's comets, a transit of Mercury, occultations of Jupiter's satellites, and new accurate estimates of the distances to the Moon and the Sun. Most crucially he made careful repeated observations of the zenith distance of the star, Alpha Centauri, with a view to measuring its distance by parallax. In his absence, Edinburgh's Calton Hill Observatory had run into trouble when the Astronomical Institution went bankrupt.

However, with the backing of Brisbane and others in 1834, he became Edinburgh's Regius Professor and the 1st Astronomer Royal for Scotland. In his 10-year tenure, he accumulated some 60,000 observations and published the first five volumes of *Edinburgh Observations*, while continuing careful analysis of his Cape data concerning Alpha Centauri, announcing its distance to the Royal Astronomical Society

Figure 5.5.1 Thomas Henderson IoP Blue Plaque: unveiling 8 October 2014 at 1 Hillside Crescent (Henderson's former home) by Professor JC Brown (10th AR for Scotland) and Dr David Gavine (author of *Astronomy in Scotland 1745–1900*)
(D McCalman (Edinburgh))

inJanuary 1839. He was the first person to obtain data that clearly showed the vast cosmic distances even to nearby stars but his slow careful data analysis meant he was not quite the first to publish a stellar distance result. Friedrich Bessel and Friedrich Struve announced parallactic distances for 61 Cygni 2 and for Vega respectively a few months earlier, though their data was gathered later than Henderson's.

Henderson had a heart condition and his general health was poor, bringing him to an early death in 1844, two years after his wife died following the birth of their only child. The final straw was likely the daily climb up Calton Hill from his home in Hillside Crescent. The Institute of Physics mounted a blue plaque to him there in 2015 (Figure 5.5.1). Calton Hill Observatory also bears a commemorative inscription and he is buried in Edinburgh Greyfriars Churchyard.

5.6 John Pringle Nichol (1804–59)

John Pringle Nichol was born near Brechin and attended King's College, Aberdeen. He went on to become a teacher, a headmaster and a writer on political economy, and beat Thomas Carlyle to become 5th appointee to the Glasgow Regius Chair of Astronomy (1836–59). He had wide-ranging interests, conducting research into the nebular hypothesis for the origin of the Solar System and the distribution of cosmic matter, but his greatest passions were more philosophical and educational in nature. Historian David Murray said that he was 'in some respects one of the most remarkable men who ever held a Chair at the University'.

He became famous not only for his inspiring lectures to students but for those he delivered to huge audiences at public meetings in and far beyond the city, making him a Carl Sagan of his day. He was also a prolific author, his famed popular *Architecture of the Heavens* reaching seven editions, while his *A Cyclopaedia of the Physical Sciences* drew wide acclaim for its width and thoroughness. George Eliot was a great Nichol fan, saying that his books let her 'Behold an infinity of floating worlds,/ Divide the crystal waves of ether pure,/ In endless voyage without port'.

He assisted with teaching in both Natural History and Natural Philosophy, allegedly inspiring the young William Thomson (later Lord Kelvin) and was instrumental in the creation of a new observatory at Horselethill for the Astronomical Institution of Glasgow in 1841. Nichol's rather casual attitude to financial matters led the Institution into financial difficulties and to the University's acquisition of the Observatory in 1845.

Nichol played an important role in improving provision of time-keeping for the City of Glasgow. He corresponded widely with John Stuart Mill and other luminaries in the worlds of philosophy, economics and literature. He conducted renowned

Figure 5.6.1 John Pringle Nichol.
(Glasgow Archives & Special Collections (Refs GN 248: UP1/258/1, UP2/26/1, and ACCN 4191/1/1))

lively arguments about cosmology with Edgar Allan Poe, who was quite knowledge-able about the topic, describing the Universe as 'a poem with a plot', in fact the 'most sublime poem with God as the plot'. Latterly Nichol was nominated for the Chair of Political Economy at the Collège de France, although he was by that time too ill to take the post, battling with opiate addiction that had arisen from prescribed medication.

5.7 Charles Piazzi Smyth (1819–1900)

Charles Piazzi Smyth was born in Italy and moved to Bedford, where his father equipped an observatory and gave him his first lessons in astronomy. At the age of 16, he became an assistant at the Cape of Good Hope Observatory, where he ob-served a Comet Halley apparition and the Great Comet of 1843, and did work on meridian transits. From 1846 to 1888, he was 2nd Astronomer Royal for Scotland and Edinburgh Regius Professor. Soon after his appointment, the observatory came under Her Majesty's Treasury control and suffered prolonged under-funding (surprise, surprise), resulting in much of his notable work being done elsewhere. His research

Figure 5.7.1 Charles Piazzi Smyth, Portrait by John Faed
(Royal Society of Edinburgh)

Figure 5.7.2 The author with the late Sergeant Tam 'The Gun' MacKay Calton Hill Time Ball Tower visible top right); Tam was incredibly popular with the countless castle visitors
(Dr MI Brown, site permission courtesy of the MoD)

included solar, auroral and zodiacal light spectroscopy, and rain-band weather forecasting. His major practical civic contribution was to establish the Calton Hill Time Ball (1852) as a visual time indicator to the Port of Leith, synchronised to a time signal from the Observatory. In 1861, for foggy days, he augmented the visual Time Ball signal with the bang from Edinburgh Castle's One O'Clock Gun, the daily firing of which is one of the city's major tourist attractions.

His most important and widely acclaimed contribution to astronomy was to be the first astronomer to heed and act on what Newton had written 250 years earlier in his *Opticks*:

[telescopes]... cannot be so formed as to take away that confusion of the Rays which arises from the Tremors of the Atmosphere. The only Remedy is a most serene and quiet Air, such as may perhaps be found on the tops of the highest Mountains above the grosser Clouds.

Piazzi Smyth's frustration with the frequent clouds and the poor stability and transparency of the Edinburgh skies drove him to raise government funds for an expedition to Tenerife in 1856, as well as for return visits. This turned out to be a turning point in the location of observatories worldwide, vindicating Newton's prediction, since Piazzi Smyth's Tenerife experiments proved the vastly superior performance of telescopes at high altitude. The world's largest telescope at that time (1845–1917) was the Earl of Ross's Leviathan (6 feet ~ 1.8 m aperture) – which can be seen in Figure 4.1.4. It was located in a boggy, misty site near Birr, Ireland, but nevertheless allowed the discovery of the spiral structure of some nebulae (galaxies – Figures 4.1.5 and 4.1.6). Eventually Caltech's Mount Wilson (2.5 m) and Mount Palomar (5 m) telescopes and most subsequent large telescopes were built in locations that recognised the need for high altitude sites (Figure 5.7.3). One exception, as late as 1967, was the UK's installation of their 2.5 m (Isaac Newton) telescope on a damp sea-level site in Sussex, later moved at great expense to La Palma in the Canaries – ironically Piazzi Smyth's original high-altitude stamping ground.

Another of Piazzi Smyth's many pioneering interests was photography. This resulted

Figure 5.7.3 ESO VLT Observatory on Mount Paranal Atacama Desert with the Milky Way and laser artificial
star beam for adaptive optics atmospheric seeing correction
(Hüdepohl (atacamaphoto.com)/ESO (cdn.eso.org/images/publicationjpg/D5C1091-CC.jpg))

in the popular *Teneriffe, an Astronomer's Experiment,* a version of the expedition report that included the very first stereoscopic photographs viewable in 3D without requiring them to be removed them from the book. He is also acknowledged for his fine artwork, which included a painting of the Great Comet of 1843, which is now displayed in the Royal Observatory Greenwich. His equally multi-talented wife, Jessica (née Duncan), was a geologist and expert photographer with a passion for science, who had been on many scientific expeditions prior to her marriage at age 43. She played a major part in Charles's expeditions to Tenerife, and later to Egypt.

The ageing Smyth's eccentric obsession with Pyramidology theories (even his gravestone is a pyramid) led him into some historic disrepute, although he had made very accurate and valuable surveys of Giza. We touch briefly on the subject of Pyramidology in Section 6.1.

Piazzi

i) Squaring the Circle

Five-year tint oan Pyramidology haivers;
Gauging Cheop's maisterpiece
Wi measurin rod an plumb-line,
Tae mak some fancy fit the facts.

Meldin the starns tae gree wi yer equations;
Draconis skinklin doun upon the threshold,
The Pleiades in polar opposition,
The Milky Way birls roond in acquiescence,

Tae merk the fore-end year o its new biggin.
The perimeter tae ye say maun kythe,
A cubit fir ilka day o the year,
An ither abracadabra's wir proclaimed.

James Young Simpson speirt quizzically at,
Yer sums, an syne produced a hat,
Tae cheers frae the crowd at the RSE,
Solemnly, wi care, computes its brim.

Hauf a cubit wide, he ettled,
An metrological maitters settled;
Mockingly he measuirt oot the Earth,
While you wi hingin lip taen in the farce.

Aa yer lear an mense hud taen the road,
An years ye spent in shorin up this keech,
While Herschel wycely sat upon the fence;
Nae faur kent fame in makkin sense o nonsense.

Yet aye-an-oan thair's thaim wha vainly sairch,
An scan yer eident notes fir dernit truths,
Wha scart their heids an heize thaim tae the heivins,
Lang-heidit, yet they sift throu dross fir diamonds.

ii) Abune the Cloods

Abune the cloods yer legacy staunds,
Whaur vultures swoop an soar,
Amang the Andean heichts;
Unblinkin een sairchin spunks o life.
Faur ablow thaim ranked in firm array,
The silvered Alma Observatory dishes;
Ilk ane centred oan some faur awa spot,
Gaitherin jigsaw pieces o cosmic lear.

Blin as mythic Tiresias,
They eident glower intil the void,
Till syne the void blents back,
A flash o wittins hained in hard-drives,
Anither clue anent the eternal puzzle.

Unsated dreamer, yer drouth wis aye unplumbed,
E'en tho ye ettled tae measuir the warld,
At the hinnerend, things didnae add up.
This ae faur-sichtit dream taen siccar ruit,
Tae pierce thae mirksome, dowie skies o Embra;
Borne awa bi Titania tae Tenerife,
Frae Alta Vista's peak ye glimpsed the future.

Piazzi Smyth, thae veesions aince ye hud
Nou glisk an glent in steel an bress,
An geck frae domes atop these fremmit mountains.
They peer intent as ony gled,
Ettlin yet tae pruive yer calculations,
An speir fir signs mang ither distant airts...

5.8 James Clerk Maxwell (1831–79)

Maxwell was born in Edinburgh and is widely believed to be the most brilliant Scottish physicist ever, in the same league as Newton and Einstein. Among tributes to him from major names in science are these:

'The special theory of relativity owes its origins to Maxwell's equations; the work of James Clerk Maxwell changed the world forever.'—Albert Einstein

'He achieved greatness unequalled.'—Max Planck

'[In] a long view of the history of mankind... there can be little doubt that the most significant event of the 19th century will be judged as Maxwell's laws of electrodynamics. The American Civil War will pale into provincial insignificance in comparison.'—Richard Feynman

'Maxwell's equations have had a greater impact on human history than any ten presidents.'—Carl Sagan

Reasons that he is not better known publicly may include the sheer number and diversity of his profound discoveries. This makes it hard to give him a single label as people like to do (for example, Newton = gravity and Einstein = relativity, though they both did lots more). Only one piece of Maxwell's work – that concerning Saturn's rings – was explicitly astronomical, but so much of his physics work is vital to astrophysics that we summarise its highlights here. For more, watch 'A Sense of Wonder' at www.youtube.com/watch?v=ANIkxDm8bF4, narrated by Rab Wilson. Maxwell developed a very early curiosity about the workings of everything and is widely quoted for the words in which he expressed this as a child: 'Show me how it doos'; 'What's the go o that?'; 'But what's the particular go o' that?'

Among his numerous achievements, where he showed the world what the go is o many things, are:

- His 1857 Adams Prize essay on the stability of the motion of Saturn's Rings, in which he proved rigorously for the first time that the rings cannot be a continuous solid or liquid but must comprise an indefinite number of unconnected solid or liquid particles in individual gravitational orbits.
- His key 1860 paper, 'Illustrations of the Dynamical Theory of Gasses', in which

he considered gases as comprising numerous well-separated particles moving freely but undergoing random velocity changing collisions with each other. He proved that if all the particles and their kinetic energy are sealed in a box and left to interact, they will settle down to have uniform density, with the fraction of particles lying in different speed ranges given by a very simple equation (the Maxwell speed distribution) depending solely on the mean particle energy (temperature). This equation proves to be vital in predicting how fast planetary atmospheres leak into space (their fastest atoms exceeding gravitational escape speed) and even how stars slowly leak away from globular star clusters (Section 3.3, Figure 3.3.6) where we think of the cluster as a gas cloud comprising massive stars as the particles. He also posed the paradox of Maxwell's demon, in which a tiny demon operates a small door in a wall dividing a box of hot gas, the demon only allowing through though the door fast atoms approaching from the left and only slow ones from the right. In this way, he can split the atoms of a warm gas into hot and cold components, which you could then use to run a heat engine. The paradox is that this violates the Second Law of Thermodynamics, unless you can prove that the demon has to do more work moving the door than you can get out of the heat engine. Discussion of that issue is ongoing even today: www.wikipedia.org/wiki/Maxwell%27s_demon.

• A series of papers from 1861–65 that brilliantly synthesised knowledge (from Gauss, Coulomb, Ampere, Faraday and others) about static and time varying electric and magnetic fields, and devising an extra term (Maxwell's Displacement Current) not then known experimentally, but needed for electric charge to be conserved. These led him to the world changing conclusions that 'electric and magnetic effects travel at the speed of light… light consists in the transverse undulations of the same medium which is the cause of electric and magnetic phenomena… light and heat are merely forms of such waves of different wave frequencies' (now known to include everything from gamma rays through x- and uv-radiation, to microwaves and VLF radio waves).

The details of the famed Maxwell's Equations, and of diagrams of the Electromagnetic (EM) Waves they predict, are beyond the general level and aims of this book. However, we think all readers may simply enjoy seeing how braw they look, so we include them here in Figure 5.8.3, which shows the four equations and a propagating EM wave. These feature the variations in space and in time of the strengths and directions of the electric (E) and magnetic force (B) vectors vary and how they are related to the strengths of the electrical charge density (ρ), and current (J) creating them.

Figure 5.8.1 James Clerk Maxwell, Portrait by Reginald Henry Campbell
(Royal Society of Edinburgh)

(D is closely related to E). λ and *c* are the wavelength and propagation (light) speed of the waves. Maxwell's theory showed that, in an EM wave, the fields E and B oscillate perpendicular to each other and to the direction of the wave propagation at light speed *c* – a universal property for waves of any value of λ from ultra-high energy gamma-rays to very low frequency radio.

These realisations, which also contain the key facets of (Einstein's) Special Relativity, have had huge ramifications in modern everyday technology such as telecommunications and lasers. Since almost all of astronomy depends on the flight of photons of light and other wavelengths across the cosmos to bring us information on its contents, Maxwell's theory of electromagnetic radiation is central to astronomy. In particular, Einstein's theory of relativity shows that particles of zero rest mass, like photons of electromagnetic radiation and gravitons of gravitational radiation, which travel at the speed of light, experience zero flight time between being emitted and detected. Hence, apart from reddening of their wavelengths by cosmological stretching of the space they are crossing, they reach us with pristine information about their source.

Maxwell solved key experimental and conceptual problems concerning colour, showing how it can be created by a suitable mix of three others. This laid the basis of all modern colour synthesis systems like RGB and of colour photography. It is commonly stated that Maxwell created the world's first colour photograph but it seems this may be a case of Chinese whispers distorting reports. The fact seems to be that he discovered by thought experiment how colour images and photographs could be created by mixing of grey scale images taken through three different colour filters. This brilliant concept paved the way for the idea soon to be turned, in the hands of contemporary photographic experts, into the world's first real colour photographs, starting in 1861 with Thomas Sutton's projection of a full colour image of a multi-coloured (perhaps tartan) ribbon – an image sometimes attributed, seemingly wrongly, to Maxwell.

Figure 5.8.2 James Clerk Maxwell aged seven
(Cavendish Laboratory, University of Cambridge)

$$\nabla \cdot \mathbf{D} = \rho$$
$$\nabla \cdot \mathbf{B} = 0$$
$$\nabla \times \mathbf{E} = -\frac{\partial \mathbf{B}}{\partial t}$$
$$\nabla \times \mathbf{H} = \mathbf{J} + \frac{\partial \mathbf{D}}{\partial t}$$

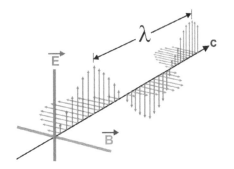

Figure 5.8.3 Maxwell's electromagnetism equations and waves
(www.pixabay.com/en/electromagnetic-waves-wave-length-1526374/)

Figure 5.8.4 James Clerk Maxwell, statue in Edinburgh by Alexander Stoddart
(Royal Society of Edinburgh)

The story of Maxwell's brilliant academic career, star-studded with accolades, has been written about widely elsewhere but the fact of his election to the Royal Societies of Edinburgh at age 25, and of London at age 29, alone speaks volumes. His personal life story as a colourful caring individual who was a lover and writer of poetry and a staunch Presbyterian is equally interesting. Among the many recent tributes paid to him are: the naming of the maxwell (M) as a unit of magnetic flux in 1930; the naming of the sub-mm-wave 15 m James Clerk Maxwell Telescope on Mauna Kea Hawaii; the statue erected to him in George Street, Edinburgh, in 2008; and the 2015 International Year of Light, in which numerous events across the world marked the 150th anniversary of Maxwell's theory of electromagnetic waves. However, the tributes paid to him by other science greats – including those quoted at the beginning of this section – represent the finest of all monuments to his brilliance.

Daftie

'From the long view of this history of mankind - seen from, say, 10,000 years from now – there can be little doubt that the most significant event of the 19th century will be judged as Maxwell's discovery of the laws of electromagnetism.'—Richard Feynman

'Whit's the go o that?!' he'd speir,
Whit wey the burn wid gang it's wimplin coorse...

'Whit's the go o that?!' he'd speir,
Whit wey the key wid birl tae lowse the lock...

'Daftie! Daftie! Daftie!' they wid screich,
Anent his hame-made shune an guid braid claith,

Thae Embra Academy unco guid,
Wha laucht at his reuch Gallovidian brogue,

Blately he wid smile an turn awa,
An taen the slings an arras o their snash.

Syne then as ane o Trinity's Apostles,
Nae fir him the *'unexamined life'*!

But whit wis the go o the Universe,
The rules an laws that govern ower aathing.

Glowerin throu the prism o Glenlair,
He'd glaum the pouer o polarising licht,

Divert his mind wi toys lik spinnin tops,
Diabolo's dancin oan the edge o naethin,

Then sit bi the road tae unraivel the knot,
An kythe the mystery o Saturn's Rings.

A chuckie stane he cast intil the burn,
Chynged in his mind til waves that wid in turn,

Meld electricity wi magnetism,
An echo seelently athort the cosmos;

His intellect ootstrippin aa his peers,
Theories unifying in his mind,

Decades lang afore the wark o Einstein,
Maxwell hud jaloused the speed o licht.

Anither Scottish hero taen ower suin,
Twa mair decades whit wid he duin?

The atoms o his dust nou lig at Parton,
The Sun streams throu the windae at Glenlair,

Whaur aince he sat tae pen his *magnum opus*,
That muckle 'giant leap' in human thocht!

'Whit's the go o that?!' he'd speir,
Whit wey the burn wid gang it's wimplin coorse...

'Whit's the go o that?!' he'd speir,
Whit wey the key wid birl tae lowse the lock...

5.9 Williamina Fleming (1857–1911)

Figure 5.9.1 Williamina Fleming, Harvard College Observatory
(HCO photos, public domain
commons.wikimedia.org/wiki/File:Williamina_Paton_Stevens_Fleming_
circa_1890s.jpg)

Dundee-born Williamina Fleming became a local school teacher at age 14, but married and emigrated to Boston, US, at age 21. At age 23, she launched into an astronomy career, which is outstanding by any measure, and all the more so given the era involved and that she and her son had been left to their own resources. Having impressed EC Pickering, then Director of the Harvard College Observatory (HCO), she was hired there to do some clerical work and mathematical calculations, where she rapidly developed a new 'Pickering-Fleming' System to classify stars by their spectra, enabling inclusion in 1890 of over 10,000 stars in the Draper Catalogue of Stellar Spectra.

Fleming was then made editor of all HCO publications and hired dozens of young women in support of her stellar research, among whom was Henrietta Leavitt, whose later discoveries concerning Cepheid variable stars led to a revolution in measuring large cosmic distances (Section 4.2). Ultimately, alongside Slipher's discoveries regarding galaxy red-shifts, this was what led Edwin Hubble to 'discover' the now so-called Hubble Law of cosmic expansion and catapulted himself into the history books. At age 41, Fleming held the key post of Curator of the HCO astrophotography plate collection and in the course of her career, with no formal astronomy education, she discovered ten novae, 52 nebulae and 310 variable stars. Aged 53, a year prior to her death from pneumonia, she measured the spectra of the first-known White Dwarf stars (see Section 3.4).

Among the nebulae she discovered were that shown in Figure 5.9.2 in the huge Cygnus Loop SN remnant, properly called Fleming's *Triangular Wisp* (but aka Pickering's Triangle), as well as the famous Horsehead in Orion (Figures 3.4.1 and 3.4.3) The latter was on a photographic plate taken by astronomer WH Pickering, brother of EC Pickering. No credit was given to Fleming or to WH Pickering for these discoveries in the first *Dreyer Index Catalogue*, which attributed the entire work to EC Pickering. However, in the second *Dreyer Index Catalogue*, Fleming and her HCO colleagues were finally properly credited for their discoveries.

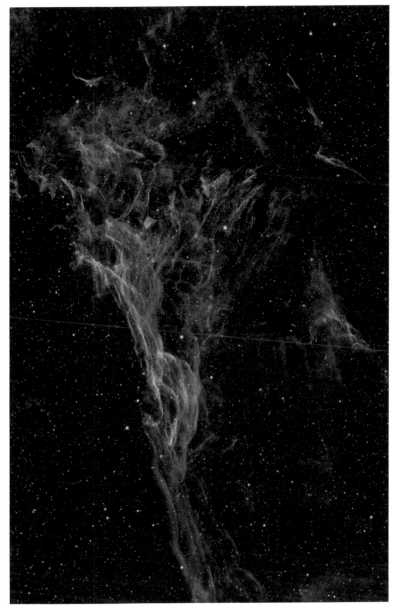

Figure 5.9.2 Fleming's Triangular Wisp, part of Cygnus Loop (AKA the Veil Nebula)
(Sarah Wager - www.swagastro.com - English Astrophotographer resident in Olocau Spain apod.nasa.gov/apod/ap171110.html)

CHAPTER 6

Human Views and Models of the Cosmos Through the Ages

In the last meenit,
O the benmaist hinner hour,
We scartit oor nems.

6.1 From Cave Art to the Late Neolithic

Cave Paintings and Other Prehistory

MUCH OF OUR material so far has been about fairly recent observation, exploration, and interpretation of oor big braw cosmos through the methodologies of modern science. In this chapter, we broaden our horizons in space, time, and discipline, looking across history at how humans have perceived, thought about, and represented oor big braw cosmos in diverse forms of art, philosophy, modelling, and engineering and architecture. In Chapter 7, we report on a few much more recent and local cosmic projects.

We first have a look at eras which are pre-historic or at least pre-science (in the modern sense). Broadly speaking, the further one looks back in time, the leaner the facts that are available. Consequently, interpretations and theories of artefacts from these eras in terms of astronomical connections are as plentiful, and sometimes outrageous, as they are fraught with rivalry, heated debate and scepticism. To avoid a barrage of outraged correspondence, we avoid here taking rigid sides or going into details of these disputes, but give a few examples of ideas out there, with some questions one might reasonably ask of them.

The earliest known human depictions of oor big braw cosmos were in the world's oldest cave paintings around 40 millennia ago, the most famous being the French Lascaux caves, dated around 20 millennia ago. Such cave painting includes features surely representing the Moon, the Sun and stars, amid numerous artworks displaying aspects of the world from humans to animals and plants etc. They are superb artworks for such an early era. Some cave investigators claim much more sophistication in cave art, such as exhibiting stellar constellations and clusters and even the lunar cycle. For example, Dr Michael Rappenglueck (Munich) suggests that: a compact set of dots above a Lascaux bull are the Pleiades (Section 3.3); a line of 29 dots are the number of days in a lunar month; a line of 13 dots are the number of days between the first visibility of a new Moon and the next full Moon. These ideas are interesting but, in cases of a single sample, they may be mere coincidences and there could be many other interpretations (eg 13 is the approximate number of lunar months in a year). However, allegedly some cave painting themes are *not* single samples, some recurring at sites widely separated in location and even in epoch. This is not the place to weigh the evidence but if such recurrences do exist and include very specific things (like the lines of 29 and of 13 dots), rather than things routinely seen like trees, bulls or crescent Moons, then even hardened sceptics should pause for thought. Most recently and extremely, a paper led by Dr Martin Sweatman of the University of Edinburgh in November 2018 presented claims that several caves across Europe show evidence in their 40,000-year-old art of awareness of the Earth's 26,000-year precession.

Scottish Neolithic Structures and Petrospheres

Figure 6.1.1 Maeshowe, outside and inside
(Charles Tait, www.charles-tait.co.uk)

Some of the world's most ancient architectural structures with clear astronomical, or at least solar, connections are to be found in the UNESCO World Heritage Neolithic complexes in Orkney – Skara Brae (village), Maeshowe (tomb), Ring of Brodgar and Stones of Stenness (standing stone circles) – and similar formations at the Newgrange complex in Ireland. All of these are even older (Skara Brae and Maeshowe are around 5,000 years old – Figure 6.1.1; Brodgar and Stenness are around 5,000 years old – Figure 6.1.2) than the conventionally accepted ages of the Stonehenge Circle and of the Pyramids (both around 4,500 years old). The large circular tomb of Maeshowe, adorned inside by much younger (about 1,000 years old) clever Viking rune graffiti, has a long passageway down which the setting Sun shines daily within about a week or so of the winter solstice. This is very like its Newgrange counterpart except that the Newgrange passage is narrower

Figure 6.1.2 Ring of Brodgar and Stones of Stennes, Orkney
(Charles Tait, www.charles-tait.co.uk)

and aims at the rising midwinter Sun for just a few days.

Somewhat similar solar alignments apply to the younger (about 4,000-year-old) Clava Cairns near Inverness (Figure 6.1.3). Standing stone circles are found widely across Scotland, some larger ones with a more elaborate layout than a circle, the most complex being the Callanish network on the Isle of Lewis (top left in Figure 3.1.12). Even the circles are not truly circular but rather ovoid with a symmetry axis sometimes north-south related. The extensive pioneering work on the stone circles was that by

Alexander Thom and collaborators, who found good evidence for cases of stone sight-lines, sometimes involving horizon features, plausibly matching solstitial sunrise and sunset, and also some extreme moonrise and moonset points.

Many features of the Scottish circles show similarities to those found in the giant Carnac stone circles in Brittany, leading Alexander Thom and others to suggest widespread communication across north European Neolithic society in building structures of considerable astronomical complexity, most likely for calendrical, agricultural and religious purposes. Today, there remains a considerable and diverging difference of opinion about the ideas of the Thom school, with conservative archaeologists and even

Figure 6.1.3 Clava Cairns
(JC Brown and HES)

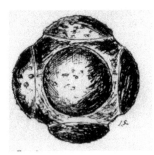

Figure 6.1.4 Petrosphere
Scotland
(commons.wikimedia.org/w/index.
php?curid=6422819)

some astro-archaeologists denying the possibility of any sophistication among such northern primitives, despite some of the evidence. At the other extreme, some new people in the field with access to novel computer data analysis methods claim even more alignments than Thom.

A very different group of early objects comprises the hundreds of small late Neolithic carved stones (petrospheres), uniquely found across what is now Scotland, but mostly north and east with ages roughly contemporary with about 2000 BC, or slightly younger than the Pyramids. Mostly around the size of a tennis ball, these carry a range of carved geometric decorations, for no known purpose other than decoration. Figure 6.1.4 shows a sketch of one petrosphere found at Jocksthorn Farm near Kilmaurs in Ayrshire. This is one of the petrospheres of the most interesting kind, since their surfaces are grooved to cover (or 'tile') the stones geodetically – ie with identical regular shapes.

This can be done in five, and only five, ways or shapes, related to the five regular solids – the 4-faced tetrahedron (triangular pyramid), 6-faced cube, 8-faced octahedron, 12-faced dodecahedron and 20-faced icosahedron. These are known today as the Platonic solids after the Greek scholar Plato who 'discovered' them two millennia *after* at least some of them were being used in Neolithic petrosphere artwork in what is now northern Scotland. The Greek interest in these five regular solids was part of their obsession with a perfect geometry-based Universe, with theories of the states of matter being linked to properties of the regular solids. This fixation delayed any real progress toward truly understanding our Solar System for 1,500 years, with even Kepler in 1596 (Section 6.2) publishing a theory linking the sizes of five nested solids to the sizes of the orbits of the five planets then known.

Afore the Dawn

Wha wis the first ah ettled?!
Lik Arthur C Clarke's ape wi his bane,
Or thon Promethean primate
Dreamin o man's rid fire.

Else, cairvin cups an rings oan stane,
When aince the penny drapt;
Mibbes we wir no oor lane.

Hou mony distant aeons passed,
O luikin doun insteid o up;
Brutes wha scartit in the muck,
An hid theirsel awa in caves.

Concernt wi anely their daily tyauve,
Tae claith an feed theirsels,
Wha's skulls held nocht but DNA,
Fir unborn anthropologists.

Lives that left nae stories,
Thir banes, repositories
Gnawed bi beasts,
Smitten wi disease.

Laid oot oan tables,
Lik calcified xylophones,
The toothless grin o Yorick,
Quite chopfallen.

Wha wis the first ah ettled?!
Aiblins this yin liggin here,
Wi mense tae leeve an extra year,
An mak thon primal lowp fir aa mankind.

That aince hud stuid up oan hint legs,
Tiltin their heid an tint in nicht,
Bricht unblinkin een that scanned the heivins,
An jaloused the starns,
Jaloused the starns!

Figure 6.1.5 Sphynx and
Giza Pyramid
(JC Brown)

The Pyramids

The Pyramids, particularly those beside the Sphynx at Giza, have surely drawn more attention – broadly called Pyramidology – than any other ancient monuments, especially in terms of cosmic connections. Among luminaries who became obsessed with Pyramids was Charles Piazzi Smyth, 2nd Astronomer Royal for Scotland (see Section 5.7). In 1865, strangely driven by the extreme ideas of John Taylor in *The Great Pyramid: Why Was It Built? And Who Built It?* (1859), Piazzi Smyth launched into the most extensive and precise measurements ever of the Giza structures. Piazzi Smyth is not alone among eminent scientists in becoming obsessed with ideas which are widely regarded by the world of science as eccentric at best.

For example, Sir Arthur Eddington, eminent researcher famed for his 1919 solar eclipse experiment testing Einstein's relativity prediction, became obsessed with the numerology of dimensionless combinations of physical constants, as did the famous quantum physics pioneer, Paul Dirac. These include the Eddington-Dirac Number – the number of protons in the Universe, which is around $10^{79\text{-}80}$. Eddington's Fundamental Theory predicted it should be exactly 136×2^{256} or about 1.57×10^{79}, and equal the square of the ratio of the electrostatic to gravitational forces between a proton and an electron. Here 256 is equal to 2^8 and 136 was the approximate value in Eddington's era of one over the atomic fine structure constant. Eddington predicted this value should be an integer, later modifying his theory to predict 137 in line with new data, but today's measurements find it to be the non-integer 137.0360286... Another example is Gerald Hawkins of Harvard-Smithsonian, well-known for his able work on meteors and his book, *Stonehenge Decoded* (1965), who latterly become obsessed by what he saw as mathematical complexities and music hidden in corn circle patterns. There was also Vincent Reddish, 8th Astronomer Royal for Scotland, who retired from an eminent career as astronomy researcher and textbook author and became obsessed with experiments on water divining and a theory of it related to Einstein's Formulation of General Relativity – *The Physics of Dowsing: Interferometry and Spin-Torsion Fields of Rotating Masses* (2003).

As regards to the Pyramids, though there continue to be heated arguments about their age, the normal paradigm today puts them around 10 per cent younger than the Neolithic stone circles or roughly contemporary with the Clava Burial Cairns and the petrospheres mentioned above. However, the Pyramids and their contents are much larger, heavier, and more elaborate and precise structures than these other artefacts. They thus continue to be an endless source of speculative theories, which are as controversial as they are marketable. This is not the place to describe these theories nor to critically review them. Suffice to say that, on the one hand, we fully accept the Pyramids individually and collectively as amazing structures even in purely civil engineering terms. On the other hand, we feel obliged to encourage great caution among aspiring Pyramidologists by giving a few examples of the kinds of beliefs and pitfalls they may encounter as they dip into the Pyramidology literature by Piazzi Smyth, Robert Bauval, Graham Hancock and others.

Firstly, concerning cosmic connections, the sides of the square bases of the Giza Pyramids do run fairly accurately east–west and north–south, so they do exhibit cardinal point orientation, as do features of some earlier Neolithic structures. However, some theories claim much more elaborate connections between the Pyramids and the sky, such as Bauval's that the not quite straight-line geometry and the differing sizes of the three Giza Pyramids reflect the geometry and brightnesses of the three stars of Orion's belt. At first sight, this claim is striking but it has been subjected to many challenges, as have other claims – for example, that a north side shaft inside the Great Pyramid was aligned with the North Celestial Pole in the epoch when the rather feeble star, Thuban, was located near there (much fainter even than the current Polaris – Section 3.2). This theory works in tandem with one that a south side shaft was aligned with where the Orion Belt stars were highest as they crossed the sky daily. Squaring the precessional epochs of these ideas and of theories about the age of the Sphynx has proved controversial.

Astronomy aside, there is a burgeoning growth in amazed claims about the number of ways in which pure dimensionless numbers like pi (π 3.14159) and the Golden Ratio arise in relationships between different lengths, and between different areas, and combinations thereof in the Great Pyramid. But all of these claims miss the point that all properties of a square pyramid are totally defined by two numbers: one fixing its absolute size (eg apex height) and one (eg slope angle X) fixing everything about its shape. So, if, for instance, the ratio of base circumference to height happened to equal or be close to a multiple of, say, π, that would define X so that every angle, and every other angle, length ratio, and area ratio would be related to π since they are all connected.

Figure 6.1.6 Sloping bus
sign and pyramid
(JC Brown)

Even wilder are claims of connections between the absolute size (eg height) of Pyramids and cosmic sizes (eg that of the Earth), distances (eg that of the Sun) and even time units. For instance, 'The height of the Pyramid in metres times the number of seconds in 12 hours is very close to the polar radius of the Earth!' Knowing this radius with precision would require the Pyramid designers to have had some form of time travel or precognition, or access to beings who did. A more bizarre example still is that Giza's latitude in degrees is 29.9792, while the speed of light is 299,792,458 m/sec. If the match of the first six digits is not coincidence, it would seem that these early builders already measured angles in degrees, had four millennias of precognitive knowledge of the metre and the speed of light, and were able to locate the Pyramid at the latitude needed to get the numbers to match.

In Pyramidology literature discussing sizes, the Giza base circumference is often quoted to at least five significant figures (sometimes seven), the former corresponding to around 1mm precision and the latter to about 10 wavelengths of light. Anyone who has seen the Pyramids will find it hard to credit that one can even define, let alone measure, lengths with anywhere near even the poorer of these precisions. Finally, to show how misleading such numerology games can be, consider the following. I read a lot about Scottish mountains and know the most legend-ridden of them all to be Ben Macdui in the Cairngorms. At 1,309 m, it is the UK's second highest mountain and is allegedly haunted by the Yeti-like Old Grey Man of Ben Macdui. I also take an interest in Pyramids and their stories, the most abundant being about the Great Pyramid of Giza, with a height 139 m. It turns out that the ratio of the height of Ben Macdui to that of Giza is equal to 3π to one part in a thousand! Secondly, the tour bus information sign on the Giza site shown in Figure 6.1.6 has the same slope as the pyramid in the background.

It is, of course, easy to make fun in this way of numerological coincidences whether they be in cosmology or in Pyramidology. So, to be devil's advocate, one might ask how much worse they are than some of the ad hoc ideas proposed to keep 'science' intact when facing observational surprises. These would include ideas like epicycles to explain planetary retrograde motion (Section 6.2) and inflation/dark energy to explain early fast/late slow cosmological acceleration (Sections 1.3 and 8.3).

In defence of science, it has to be said that scientists always strive for ideas which lead to falsifiable predictions. These are largely absent in fields like Pyramidology.

6.2 The Geometrically Perfect Cosmos of Ancient Greece and its Legacy

The scholars of ancient Greece made many remarkable advances in mathematics and logic such as Euclid's putting plane geometry on a sound logical basis of axioms and theorems. He also devised the first proof that there is no largest prime number (an integer only exactly divisible by itself and 1, such as 2,3,5,7,11). About the same time, Aristarchus devised a method to measure the distance to the Sun, though his result was about 50 times too small, and advocated a Sun-centred Solar System. Around 130 BC, Hipparchus created the first western catalogue of stars, building on earlier Babylonian work and, around 150 AD, Ptolemy published his Almagest, a treatise on the mathematics of the apparent motions of the stars and planets but one adhering to the incorrect paradigm of an Earth-centred (geocentric) Solar System which was to prevail for the next millennium and a half. One scientific reason for opposition to the heliocentric model was that, if the Earth went around the Sun, we should see annual parallactic shifts of the stars unless they are unbelievably far away. Such motions were, in fact, not successfully measured until the early 19th century by Thomas Henderson and others (Section 5.5) – because the stars *are* 'unbelievably' far away (around 200,000 times further than the Sun).

However, the geocentric dogma was also strongly underpinned by the Greek belief in mankind's special position in the Universe and their preoccupation with geometric perfection. For instance, the motion of planets around our skies exhibit the initially surprising phenomenon called retrograde motion, looping back on their tracks for a while as shown on the left of Figure 6.2.1. This part of the Figure also shows how retrograde motion actually comes about. It is simply because both the Earth and the planet are circling the Sun, and the Earth overtakes the planet during part of the orbit. However, the long-held insistence on the world being geocentric with all motion being in circles around the Earth forced astronomers to devise a complicated ad hoc scheme of epicyclic/deferent planetary motion shown on the right of Figure 6.2.1. This involved circular motion around an epicentre, which is itself in uniform circular motion round a centre, repeated to higher orders if need be to fit the data.

Models of reality were thus at that time greatly controlled by deductive reasoning from beliefs about how things should be rather than by inductive inference of how they actually were from carefully measured facts. This mode of thinking persisted in the west for around 1,500 years, contributing greatly to the slow development of modern science through our Dark Ages prior to the Copernican revolution. This was despite the parallel key advances being made by Islamic astronomers to both observations and conceptual thinking, often still overlooked in western versions of history. For example, in his book, *Mysterium Cosmographicum* (1596), even Kepler contended that the orbital radii of the six then-known planets would fit into the spaces between and around a nested set of the five Platonic solids (Figure 6.2.2).

However, Kepler moved on through careful study of detailed observations of planetary motion that had been collected during this period by Islamic astronomers during this period and later by Tycho Brahe, coming to realise the true elliptical shape of planetary orbits and the equations describing their varying rate of motion round these ellipses. Together with the telescopic discovery by Galileo of imperfections in the heavenly spheres (Sunspots, lunar craters, Saturnian rings, Jovian Moons, Venus phases), these discoveries were the ultimate drivers in the Copernican revolution, and the modern concept that scientific truth is to be found by testing falsifiable hypotheses against data. The obsession with geometry and uniformity as the basis for theories from which predictions could be made deductively gave way to a much wider ranging mathematical approach, such as the invention of the calculus by Isaac Newton and Gottfried Wilhelm Leibniz and hence to Newton's inverse square central force Law of Gravity (1687), which predicted exactly Kepler's empirical laws of elliptical motion.

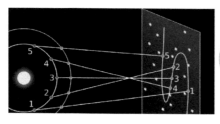 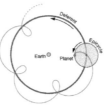 

Figure 6.2.1 Retrograde motion and epicycles
(JC Brown; Brian Brundel commons.wikimedia.org/wiki/File:Retrograde_Motion.bjb.svg)

Figure 6.2.2 Kepler's Mysterium Cosmographicum
(Kepler 1596)

The Grand Orrery

*IM Michael Bennett-Levy, restorer of The Grand Orrery
located in Dumfries House*

Here in the entrance hall o Dumfries Hoose,
Haudin sway ower Meissen or Chippendale,
Sits the cockapentie auld Grand Orrery.

Lordin it ower lesser warks o art,
His glentin gears an cogs o burnisht bress,
Defy the centuries gane syne he wis craftit.

No fir him some tea-table miscellany!
The idle crinoline clish-clash o the ladies…
His concern is o much heicher things!

Celestial orbs wha trace thair endless raikin,
Tae airts that's hyne-awa ayont oor ken,
An's cam wi bleezin portent hame agane.

Anely a marquis cuid caw *his* winding haundle,
Tae set wir Solar System whummlin roond,
Helios glentin gowden oan his axis.

In gentle minuets the spheres aa spin,
An dance their timeless dance around the heivins,
At graceful intervals they meet an pass.

His case inhauds sic miracles an ferlies,
But cantily he'll share wi ye his lear,
Kennin fine ye'll tell whit airt ye goat it!

Wha's een hae bin enriched bi this braw sicht
As whirlin throu the firmament they gang;
Nae wunner he's as proodfu as a cock!

6.3 Mechanical Models of Time, Space, and Motion

Ancient Egyptian obelisks are among the first recorded time-keeping devices and evolved into ever more sophisticated sundials. The fact that these were useless in cloudy weather, and at night, led to the use of changing stellar positions (such as rotation of the Plough around the pole) as night clocks and to the invention of various mechanical means of measuring time in any weather. The earliest of these were water clocks, possibly from around 4000 BC, in China and certainly by around 1500 BC in Babylon and Egypt. In these, the passage of time is measured by the rate of dripping or flow of water – similar to sand in an hourglass. Progressive incorporation over the years of various forms of gearing improved the accuracy of such clocks. The much more precise pendulum-regulated clock was only created by Huygens in 1656, though Galileo had noted the concept by 1583. Ever-increasing precision of mechanical clocks was attained (see, for example, the great story of John Harrison's marine-chronometers told by Dava Sobel in her bestseller, *Longitude*) through such devices as multi-metal pendulums, whose length did not change with temperature, and ever more sophisticated escapement mechanisms. Eventually, discoveries in electromagnetism led to further refinements of clock control via electro-mechanical devices. These were the pinnacle of clock technology until they were superseded after 1927 by clocks regulated by electronic quartz crystal oscillators and, ultimately, by atomic clocks using atomic vibration periods as the time reference.

Figure 6.3.1 Grand Orrery 1758 in Dumfries House
Meticulously restored by the late Michael Bennett-Levy, seen at work in this image
(reproduced by kind permission of Dumfries House, part of the Prince's Foundation, and of Michael's widow Zoë Bennett-Levy and family. A fascinating short video about the Orrery restoration, created by Kenny Caldwell of Being There Media, is available at www.youtu.be/U7gtrtKPJVY)

Figure 6.3.2 Fulton Orrery 1831
Kelvingrove Museum and Gallery, Glasgow
(Glasgow Life, Kelvingrove Museum)

In parallel with these developments in modelling the passage of time, human ingenuity also went into devices such as astrolabes for spatial measurements of the 2D angular position of objects in the sky. These are known to have been in widespread use throughout the Islamic world by the 8th century AD, while the vast collection of instruments at Jaipur, India, bears witness to the extent of early parallel Hindu studies of the sky.

3D models of the cosmos of that time – really just of the Solar System – came much later because distances were huge, poorly known, and hard to show on the same scale as the comparatively tiny sizes of the planets. In Solar System models like that built in 1776 at Kirkhill by the 11th Earl of Buchan, planet sizes were shown on a larger scale than planet distances in order for them to be visible at all. For example, with the Earth scaled down to 1cm size, the Sun would be about 1m across but, on the same scale, about 200 m away, while the distance to Neptune would be 6 km.

The same era as that which saw Buchan create his static model of the Solar System and his Kirkill Pillar predictions (Section 7.5) of far in the future planetary positions also saw the birth of modern orreries – small, table-top moving mechanical models of the Sun and planets, starting early in the 1700s AD. Two fine extant examples in Scotland are the 1758 Grand Orrery in Dumfries House (Figure 6.3.1) and the one now in Glasgow Kelvingrove Museum, made in 1832 by the self-taught John Fulton of Fenwick (Figure 6.3.2). The latter was the first to include the planet Uranus (discovered in 1781) but not Neptune (discovered in 1846), and also a number of planetary moons complete with inclined orbits. The world's oldest working orrery is the the Royal Eise Eisinga Planetarium, made in 1781 and located in Franeker, Friesland, the Netherlands. Created by Eise Eisinga and nominated for UNESCO World Heritage status, it has planets revolving around slots in the ceiling. Ancient Greece did have orrery-like models (sometimes called planetaria) of the planetary system but much less

Figure 6.3.3 (left) Meccano 'Orrery' of Jupiter's rotation and the Galilean moon orbits by Christine Cooper when aged 10
(Christine and Douglas Cooper)

Figure 6.3.4 (right) Royal Eise Eisinga Planetarium, Franeker, NL
(Erik Zachte (commons.wikimedia.org/w/index.php?curid=72367174); Bayke de Vries (commons.wikimedia.org/w/index.php?curid=34787955)

elaborate than these later ones. At the other extreme, it is well within the abilities of a motivated 10-year-old to make a simple orrery such as the working Mechano model of Jupiter and its Galilean moons by Christine Cooper (Figure 6.3.3).

However, the most amazing discovery in mechanical modelling of the Solar System and more besides is the Antikythera mechanism, an analogue computer able to predict astronomical positions and eclipses for calendar and astrological purposes. Discovered in 1902 in a Greek ship wreckage, and dated at around 100–200 BC, this small structure was studied by Mike Edmunds and Tony Freeth (Cardiff University) using x-ray tomography and high-resolution surface scanning. It was found to contain an elaborate gearing mechanism designed to predict things as diverse as eclipses, variations in the Moon's orbital speed, and the four-year cycle of the ancient Olympic Games. See www.wikipedia.org/wiki/Antikythera_mechanism for details.

6.4 Modern Planetaria and Virtual Reality Systems

While some computer, tablet and smart-phone software and apps that allow on-screen viewing of the night sky charts use the term planetarium in their name, in present day parlance, a real planetarium is a full dome projection system in which a seated audience views a projection of the upper hemisphere of vision. These comprise large fixed-dome systems like that in Glasgow Science Centre (Figure 6.4.1), which seat about 100 people, and small inflatable portables like Cosmosplanetarium, which seat about 30 on the floor, that can be taken out to schools and shopping malls and the like. (Figure 6.4.2). Planetarium domes are also increasingly used for non-astronomical projection material ranging from psychedelic full-dome graphics accompanying, for example, Pink Floyd albums, or fly-through 3D videos of buildings. Traditional opto-mechanical projectors use starballs with holes and lenses to project

Figure 6.4.1 Glasgow Science Centre Planetarium
(GSC and Martin Shields www.martinshields.com)

Figure 6.4.2 Cosmos Planetarium – mobile
planetarium in a shopping arcade
(Steven Gray www.cosmosplanetarium.co.uk)

fixed images of cosmic objects, while more recent systems use fibre-optics and the like to make the stellar images even more real – so real that it is hard to believe one is not under a real night sky. Such systems do not, however, enable changes in stellar positions so one cannot 'fly through' space as one can do in modern digital dome projection systems. The latter are also attaining ever better realism in projecting bright twinkling stellar points while simulating stunning 3D flights through the cosmos. A major current trend, driven by the games market, is towards achieving the same 3D experience at the individual level as in a dome by means of virtual reality immersion stereo goggles.

6.5 Modern Artistic Representations of the Cosmos

In parallel with the many technological advances discussed above used in displaying and modelling determinist/realist views of our big braw cosmos, artists of all disciplines have of late ventured into ever more amazing ways of presenting perceptions of it. As well as the work of painters like Lynette Cook and musicians like Herschel 36 already mentioned in our Introduction and featured in this book, there are many great cosmic artists working with diverse media such as installation and landscape art. These include the large-scale landscape work of American-born Charles Jencks inspired by cosmology and particle-physics. Examples are the *Garden of Cosmic Speculation* (Figure 6.5.1) at his Portrack home near Dumfries and the Crawick *Multiverse* (Figure 6.5.2) and a re-landscaped opencast mine near Sanquhar on the River Nith in Dumfries and Galloway.

Antony Gormley, Angel of the North creator, has explored cosmic themes such as in his *Blinding Light*. There, one entered and walked through a marquee filled with nothing but vapour and bright light in every direction, as if in the interior of a star but without the scorching heat. Others who explore the nature, shape and light of the Universe include: James Turrell, for example, in his *Cat Cairn: The Kielder Skyspace* and his *Roden Crater*, Arizona; Liliane Lijn, in her *Supercollider* and *Moonmeme*; and Gill Russell's *Sòlas* (Figure 6.5.3), which involved a 15-minute

Figure 6.5.1 *Cosmic Cascade Staircase*, Jencks, Portrack Gardens, Dumfries
(Charles Jencks)

Figure 6.5.2 *Galaxies, Crawick Multiverse*, Sanquhar
(Charles Jencks)

Figure 6.5.3 *Sòlas* Installation at Glenuig 2007–8
(Gill Russell www.gillrussell.co.uk/p/solas.html)

walk through Glenuig woods to a lovely west-facing headland site (see also Section 7.2). There, the *Sòlas* installation gathered light by day from our nearby Sun, and by night sent it back out into the furthest recesses of our night sky.

CHAPTER 7

Some Personal Adventures in Oor Big Braw Cosmos

Jock Broon tilts his lance,
Rab, his trusty fiere, his pen –
At Cosmic Windmills!

7.1 Solarigraphy

WEST SCOTLAND IS not renowned for its sunshine, but its mix of sun, cloud, mist and rain provide wonderful skyscapes, sunsets and greenery. Also, contrary to common perception, it does experience occasional arid spells. These include early summer in 2012, when livestock was parched and production slumped at the water-starved Talisker Distillery in Skye, and March/April in 2013, when wildfires abounded in northwest Scotland.

A simple and beautiful way of recording long-term weather variations is by using long-exposure pinhole camera photography called Solarigraphy (or Solargraphy). In its simplest form, this involves a piece of very low ISO speed photo-sensitive paper curved to fit round the inside of an opaque cylindrical container (for example, the plastic tubs from a 35 mm film) opposite a small hole in the wall covered by baking foil with a tiny pinhole in it. The container is mounted firmly facing in a fixed direction (usually south) and left with the pinhole opened for six months (midwinter to midsummer or vice versa). Pinhole cameras have a very wide angular field of view (up to 180 degrees so that the Sun, when visible, leaves a narrow trail on the film, tracking its diurnal path across the sky with gaps when the Sun is behind cloud and the most intense parts of tracks occuring when the Sun is highest in the sky. As the date advances from midwinter, the daily arcing track becomes wider and higher, recording progression of the seasons but also the changing weather pattern with tracks completely absent in prolonged cloudy periods and so forth. Solargraphs thus provide a continuous record of sunshine variations through the days and the months of any one half-year and between different half-year periods.

A sequence of examples of Solarigraphic weather records taken in the Skye/Lochalsh/Kintail area of northwest Scotland is shown in Figures 7.1.1–7.1.4, and in Glasgow in Figure 7.1.5. These illustrate clearly that Scotland *does* experience a lot of cloudy spells but also quite a lot of sunny spells, and that the dates/times/durations of

Top to Bottom: Figure 7.1.1 Solarigraphs, Breakish, Skye, early 2011 (JC Brown)

Figure 7.1.2 Solarigraph, Eilean Donan, late 2012 (Alex McKay, 1964–2018)

Figure 7.1.3 Solarigraph, Skye Bridge, late 2012 (JC and MI Brown, H Davies (Eilean Bàn))

these vary widely from year to year.

Figure 7.1.1 comprises two Solarigraphs from Breakish, Skye, for the first half of 2011, showing very sunny spells in March/April and late June, but a very cloudy wet April/May. One of the films shows the damage caused by water ingress through the pinhole during exceptionally heavy rain on the so-called Misty Isle. Note that these Solarigraph exposures start from midwinter, when the solar arc is lowest and successively higher arcs correspond to dates progressively nearer to midsummer.

Figures 7.1.2 and 7.1.3 comprise Solarigraphs for the second half of 2012 from two different sites, in each case alongside a regular (narrower field) camera photograph of the same views. In this case, the Solarigraph exposures started from mid-year when the Sun is highest, and successively lower arcs correspond to dates progressively nearer to midwinter. The first images (Figure 7.1.2) were created by the late great Alex McKay, former tour guide at the world famous Eilean Donan Castle just south of Skye. The castle appears in his Solarigraph and the nearest of Skye's Red Cuillin Hills is just visible on the right end of the horizon.

About 10 miles north of the castle lie Kyle and the Skye Bridge, which replaced the Isle of Skye (Kyle to Kyleakin) ferry in 1995. After a causeway section to Eilean Bàn (one-time island home of *Ring of Bright Water* author, Gavin Maxwell), the main bridge soars high over to Skye. This is seen in the two parts of Figure 7.1.3. As well as the bridge, the images show Maxwell's former home and the Eilean Bàn (Stevenson) Lighthouse in the bottom left, with Beinn na Caillich near Broadford just visible in the bottom right. Eilean Donan and Eilean Bàn are close enough together to experience quite similar weather and sunshine records, both showing (top bright arc) the brilliant prolonged sunny spell of early summer 2012 that made northwest Scotland the driest place in the UK. While much of the UK was suffering heavy rain, Skye was in drought.

On the Skye side of the bridge, an immediate left turn takes us into Kyleakin, beside the harbour of which lies the Bright Water Visitor Centre led by Hugh Davies, where there is an otter sculpture outside. Figure 7.1.4 shows a fine Solarigraph taken from the Visitor Centre by Hugh during the period of February to July 2013. It shows the otter sculpture in the centre foreground and, on the left across the harbour bay, the ruins of Castle Moil, a fine walk from Kyleakin at low tide.

Solarigraphy is, of course, not restricted to rural settings, and some fantastic cityscape results have been obtained. For instance, Figure 7.1.5 shows an exposure for the first half of 2004 shot across the River Clyde toward Pacific Quay, with Glasgow Science Centre (GSC), Glasgow Tower and IMAX. (There was no BBC building at that time.) Taken by Mario Di Maggio (then of GSC Scottish Power Space Theatre/Planetarium), it shows both primary solar tracks and their reflections in the river.

Figure 7.1.4 (above) Solarigraph, Kyleakin, early 2013
(H Davies (Eilean Bàn))
Figure 7.1.5 (right) Solarigraph across River Clyde to
Glasgow Science Centre
(Mario Di Maggio)

Like numerous Solarigraphers, all of us involved here are grateful to Tarja Trygg of Helsinki, leader of an international Solarigraphy collaboration (www.solargraphy.com) for providing us with pinhole cameras and for processing our results.

7.2 Two New Constellations for Scotland

The year 2009 was not only designated as the Scots Year of Homecoming/Burns Year to mark the 250th Anniversary of Robert Burns' birth, but also the International Year of

Figure 7.2.1
Logo of the International
Year of Astronomy
(IYA)

Astronomy (IYA) to mark the 300th Anniversary of Galileo's use of the telescope and discovery of Jupiter's moons, Saturn's rings etc.

Burns is not without other space connections. These include the 217 orbits of the Earth (5.7 million miles) spent aboard the ISS in 2010 by a miniature (one inch high) book of Burns' poetry. The book was taken aboard NASA Shuttle STS-130 *Endeavour's* to the ISS by British born Astronaut, Nick Patrick, on behalf of ten young Scots participants in the Scottish Space School (University of Strathclyde), who delivered it to him when visiting NASA JSFC in Houston. It was given by Nick, after his flight, to the then First Minister for Scotland, Alex Salmond (see also Section 5.0, Figure 5.0.1). Additionally, it has been suggested that Burns' 'Auld Lang Syne' likely holds a world record for the longest continuous singing of a song. The reason is that it is sung at local midnight on Hogmanay by Scots and Scotiaphiles on most, if not all, lines of longitude around the globe, excluding longitude regions such as oceans, deserts, Arctic and Antarctic regions devoid of people (but counting ships etc!).

From late 2008 through the end of 2009, vast numbers took place worldwide in extremely diverse IYA-celebratory events, organised for and by people of all ages and backgrounds. One of many in the UK was the country-wide New Constellation for Scotland project led by artist Gill Russell in conjunction with me. This built on Gill's very successful earlier Highland constellation project for Scotland's year of Highland Culture Constellation project (2007). Seven schools participated in a series of workshops involving a planetarium, astronomy, creative writing, and art, leading to choices of seven different bright 'circumpolar stars' – stars near enough to Polaris to be visible all year round.

Even the brightest stars are so far away that their light takes years, or even thousands of years, to reach us, so we are looking back in time. In each school, the pupils researched the distance of their chosen star – hence the date at which the light we see now left them – and so could explore what was happening in their community back then. Armed with all this exciting astronomical and historical knowledge, along with their creative writing and art experiences, every pupil involved was invited to design, draw and paint a new constellation that included these seven stars amongst its features. Of many interesting submissions, the best and most apt was judged to be the Jumping Fish (Lasg a' Leum) by Alexander of Ullapool. Gill subsequently created three project-related outdoor installations at Glenuig, opened with a torchlight procession and entertainments in November 2007. A book of the project created by the schools and including new work by Linda Cracknell, and a Star Jewellery collection, formed part of the Highland 2007 exhibition opened at the Scottish Parliament on 5 December 2007.

The 2009 IYA Scottish Constellation Project followed similar lines except that the nine schools selected were spread all over Scotland, the workshops included some magic demonstrations and lessons (as well as astronomy content) from me, and the method of choosing stars was quite different. In this case for each school, we had identified a historically important astronomy-related site in their part of the country, ranging from the 74-year-old Dundee Mills Observatory to the over 5,000-year-old Maeshowe in Orkney. The pupils had to research and choose between several bright and interesting cosmic objects anywhere in the sky but located at about the distance corresponding to the age of their site. For example, one school in that area of Scotland had to choose

Figure 7.2.2 Jumping Fish
(Lasg a' Leum)
(Gill Russell)

Figure 7.2.3 Wee Timorous Beastie
(Photo JC Brown; original art, Laura D, Dalmeny School)

a star or object at about 74 LY so that the light we saw from it then left it when Mills Observatory was being built.

As in the Highland Constellation case, this led to the challenge of designing and painting a constellation around the complete set of stars, in this case a rather big constellation since the chosen stars were spread across the sky. On 30 June 2009, in the Glasgow Science Centre Planetarium, four out of 200 entries from across Scotland were presented prizes by Liz Lochhead, Scots Makar at that time. The winning submission was *Wee Timorous Beastie* by Laura from Dalmeny Primary 7, inspired by the fact that 2009 was Burns Year as well as IYA. The other prizes went to Laura at Lauder Primary, Abdur at Glasgow Glendale, and Ruby at Broadford.

Yondermaist

Luikin at the starns,
We traivel back in time;
A journey ayont memory...

Beams o licht frae Betelgeuse
Began their antrin odyssey,
As Galileo trained his scope
Oan Jupiter's munes;
The human key that lowsed the lock
O this universal Pandora's box.

Poetically nemmed Ras Elased Australis
Heralded the birth o Burns;
Astrologers ettle it kythes the makar's gifts;
Pouer o language,
Glib-gabbit genius o gash expression...

Ither licht traivels a billion year,
Blinks frae the skinklin welkin o nicht,
Maks us smile, coost oor een tae the lift
An mak a weesh;
Starnlicht wis aye a hamelie thing.

Nae wunner oor weans maun blithely juggle,
Horatio's philosophy an unkent dreams,
An heeze *mair* constellations intil heiven.
Canty tae gang these nemless roads,
Reflectin back whaur man hus nevir trod...

Measuirt oot in the impossible,
The distance frae their schuil tae Neverland;

Dalmeny tae Menkar,
Lauder tae Mirach,
Glendale tae Bellatrix,
Coupar Angus tae Aldeberan…

Kirstenin ilk new constellation;
'Wee Sleekit Beastie'
'Mermida'
'Lost in the Jungle'
'Seraurora, the Whisperer of Lights'

Daein whit we've ayewis duin;
Takkin whit's ayont oor reach,
Owersettin the yondermaist,
Til somethin we can grasp…

7.3 The Astronomer Royal for Scotland – Tales of a Coat of Arms

The post of Astronomer Royal for Scotland (ARfs) was created in 1834 at the Royal Observatory Edinburgh (ROE) on Calton Hill to widen the British/UK national base for astronomy including methods of time-keeping and navigation. This extended the time-keeping network and augmented the work of the Astronomer Royal (AR) post at the Royal Observatory Greenwich (ROG), created in 1675, prior to the UK Union. The first two ARfs incumbents were Thomas Henderson

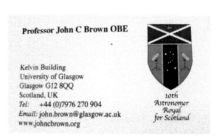

Figure 7.3.1 ARfS Business Card
(JC Brown)

(Section 5.5) and Charles Piazzi Smyth (Section 5.7). From 1834, the ARfs post was tied to Directorship of the ROE and to the University of Edinburgh's Regius Chair of Astronomy, just as the AR post was tied to the Directorship of the ROG from 1675 for around 300 years.

Over the period of 1948–57, the scientific work and staff of ROG was moved to Herstmonceux Castle in Sussex, a site better (though not ideally) suited to observing than the now highly urbanised ROG area, and near the new University of Sussex Astronomy Department. ROG was joined there in 1967 by the new 98-inch Isaac Newton Telescope (INT) for a while (see Section 5.7) but it was moved to a sensible high-altitude site in La Palma.

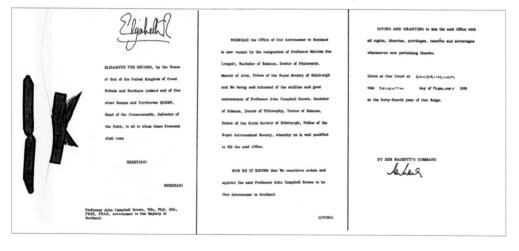

Figure 7.3.2 ARfS Royal Warrant 1995
(JC Brown)

Figure 7.3.3 UKATC opening 1998 and signed 1999 eclipse first day cover
(Dr MI Brown)

In 1990, there followed another ROG upheaval with its move to Cambridge Observatory alongside the Institute of Astronomy, but it only survived there until its final closure in 1998 after 323 years as a research entity. This followed a contentious and divisive period of political (Observatory Wars) pressure to streamline and rationalise the work of the two Royal Observatories, ending in the relocation of some ROG staff to ROE and the opening there in 1998 of the new UK Astronomy Technology Centre. This built on ROE's previously strong record in instrumentation and integrating it with some transferred from ROG.

When 11th AR Sir Richard Woolley retired in 1971, Professor Margaret Burbidge was appointed as ROG Director, and the AR post was decoupled from ROG to become an Honorary appointment for whoever was deemed by Downing Street and the Regent to be best suited to represent and promote UK astronomy. The first appointee to this new role in 1972 was Professor Sir Martin Ryle of Mullard Radio Astronomy Observatory at Cambridge who was awarded the Nobel Prize in Physics with Anthony Hewish in 1974, the first such award to recognise astronomy research. The departure in 1993 of 9th ARfs Regius Professor Malcom Longair to head the Cambridge Cavendish Lab and the developments at ROE/UKATC led to a similar decoupling of the ARfs from its tie to the ROE Directorship and Edinburgh Chair, so that it became an honorary appointment similar to the AR arrangement since 1972. On 7 February 1995, as the then Astrophysics Chair holder in Glasgow, I was appointed by HM the Queen as the 10th ARfs – the third Scots-born and the first not to be Edinburgh-based (see Figure 7.3.2).

At the time of the ROE UKATC opening in 1998, the four out of ten ARfss still living are shown in Figure 7.3.3 – Bruck 7th added digitally, Reddish 8th, Longair 9th and Brown 10th ARfs, alongside a Royal Mail First Day Cover issued for the 1999 solar eclipse, signed by all four.

Once I got over the shock that a dissident like me had been offered and accepted the ARfs post, I found myself having to reply to some of my witty well-wishers who asked whether there was an annual salary, or at least something in kind like a case of good malt, a basket of lampreys, or perhaps a starry cloak and pointy hat like Ali Bongo's, my being a semi-pro magician. Sadly, the answer to these was no, the letter from the Secretary of State saying, 'this post carries no official duties, nor any remuneration'. On the other hand, the Royal Warrant signed by ER II states '...shall be entitled to such rights, benefits, privileges and appurtenances as attach thereto'. Mind you, my wife and I have as yet to find any such though we still feel we should at least have sheep-grazing rights somewhere on Blackford Hill!

I was also surprised to learn that the ARfS post had no official Coat of Arms and, having an arty streak, decided to try and create some and to get them approved, a process which proved trickier than expected but also fun, especially visiting and working with the Lord Lyon, King of Arms – at that time Sir Malcolm Innes of DeKnight – on the design. He and his staff in their Edinburgh Registry Place Office were a delight to work with, guiding me through the process and rules involved. The first hurdle such an application faces is to persuade the Lord Lyon's office that you have grounds for applying to have officially recognised Armorial Bearings (Coat of Arms) rather than just fancying designing your own flag to wave. Fortunately, having Royal Warrants

as Glasgow Regius Chair and as Astronomer Royal for Scotland got me past this hurdle and I then *only* had to create a design which conformed to the Rules of Heraldry, met with the Lord Lyon's legal and aesthetic approval, and pay the fees due. The whole process from first enquiry to issue of the final document and artwork took about 14 months, mainly because all concerned had umpteen other things on their plates.

The graphics work in the design was my own, progressing from hand-drawn scribbles to Apple Superpaint on a Mac SE. From the beginning, I was keen to involve several elements in the design: the constellation of Orion as my favourite since age 10; something Scottish like the Saltire or Lion Rampant; something royal like a crown or Lion; and something representing astronomy hardware. These wishes survived through to the final

Figure 7.3.4 Progressive ARfS arms designs
(JC Brown)

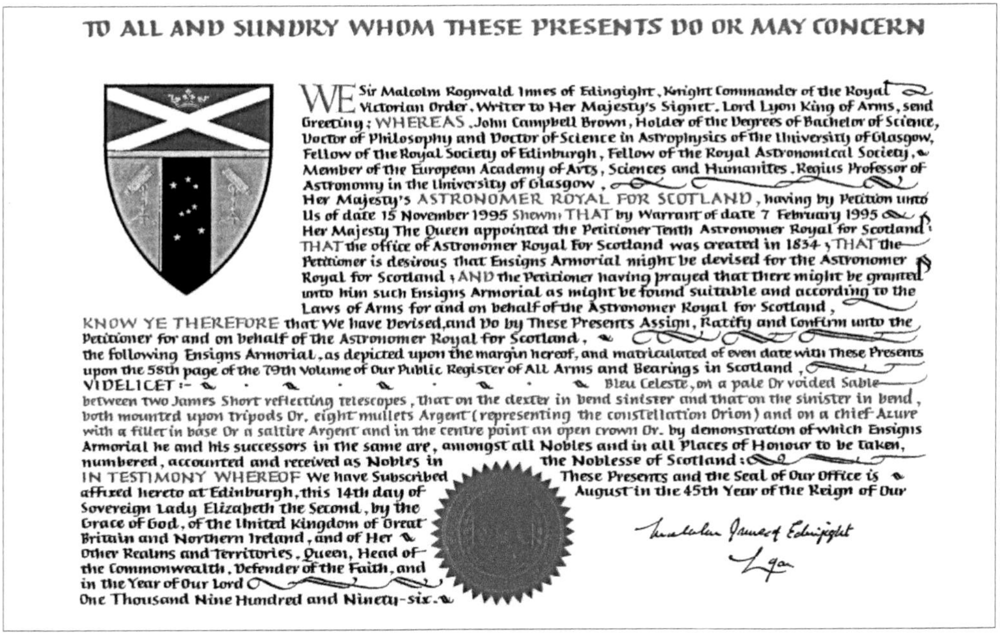

TO ALL AND SUNDRY WHOM THESE PRESENTS DO OR MAY CONCERN

WE Sir Malcolm Rognvald Innes of Edingight, Knight Commander of the Royal Victorian Order, Writer to Her Majesty's Signet, Lord Lyon King of Arms, send Greeting; WHEREAS, John Campbell Brown, Holder of the Degrees of Bachelor of Science, Doctor of Philosophy and Doctor of Science in Astrophysics of the University of Glasgow, Fellow of the Royal Society of Edinburgh, Fellow of the Royal Astronomical Society, Member of the European Academy of Arts, Sciences and Humanities, Regius Professor of Astronomy in the University of Glasgow, Her Majesty's ASTRONOMER ROYAL FOR SCOTLAND, having by Petition unto Us of date 15 November 1995 Shown; THAT by Warrant of date 7 February 1995 Her Majesty The Queen appointed the Petitioner Tenth Astronomer Royal for Scotland; THAT the office of Astronomer Royal for Scotland was created in 1834; THAT the Petitioner is desirous that Ensigns Armorial might be devised for the Astronomer Royal for Scotland; AND the Petitioner having prayed that there might be granted unto him such Ensigns Armorial as might be found suitable and according to the Laws of Arms for and on behalf of the Astronomer Royal for Scotland, KNOW YE THEREFORE that We have Devised, and Do by These Presents Assign, Ratify and Confirm unto the Petitioner for and on behalf of the Astronomer Royal for Scotland, the following Ensigns Armorial, as depicted upon the margin hereof, and matriculated of even date with These Presents upon the 58th page of the 79th Volume of Our Public Register of All Arms and Bearings in Scotland, VIDELICET:- Bleu Celeste, on a pale Or voided Sable between two James Short reflecting telescopes, that on the dexter in bend sinister and that on the sinister in bend, both mounted upon tripods Or, eight mullets Argent (representing the constellation Orion) and on a chief Azure with a fillet in base Or a saltire Argent and in the centre point an open crown Or. by demonstration of which Ensigns Armorial he and his successors in the same are, amongst all Nobles and in all Places of Honour to be taken, numbered, accounted and received as Nobles in the Noblesse of Scotland: IN TESTIMONY WHEREOF We have Subscribed These Presents and the Seal of Our Office is affixed hereto at Edinburgh, this 14th day of August in the 45th Year of the Reign of Our Sovereign Lady Elizabeth the Second, by the Grace of God, of the United Kingdom of Great Britain and Northern Ireland, and of Her Other Realms and Territories, Queen, Head of the Commonwealth, Defender of the Faith, and in the Year of Our Lord One Thousand Nine Hundred and Ninety-six.

Figure 7.3.5 Final ARfS arms certificate
(JC Brown)

version but not before quite a few fell by the wayside following various rulings and advice from *The Lyon*. The first draft, using telescope domes as astronomy hardware, and involving both the Saltire and the Lion Rampant, fell afoul of the fact that use of the Lion Rampant requires personal consent of the Queen as it is her insignia in Scotland. While such consent might well have been given, The Lyon advised it could take quite some time especially as, in 1996 there was major re-organisation of local government in Scotland, creating a queue for approval of letterhead logo redesigns. This was also happening in parallel with a trend among universities and colleges to buy the idea that their fortunes could be transformed by paying enough to logo redesign consultants.

In the next attempt the Lion Rampant was replaced by crowns as the 'royal' feature and the domes with a schematic of a modern telescope. I thought it interesting but too lop-sided for Arms. My third try placed the crown above the top saltire panel and a symmetric pair of such telescopes looking above Orion toward the crown. My efforts were improving but the telescope style was amateur rather than professional and not really apt for the ARfS insignia. So, I brought back the domes – and believed we were nearly there – even The Lyon agreed! However, being an old hand at this, he would not approve it till he applied his *fresh-eye test* – putting it away and ignoring it for some weeks then checking first impressions on bringing it back out.

His verdict: 'the domes will have to go – you and I know what they are and we like them – but the punters will see them as swing-top pedal bins. Can you find an alternative?' – and he was dead right.

So, it was a case of back on with the thinking cap for a while until the final eureka moment when I found a domes solution on which we both happily agreed. This was to introduce a very apt historic element by replacing the domes with schematic images of early telescopes made by the renowned Edinburgh mathematician and telescope maker James Short (1710–68).

With hindsight, I also liked the fact that, in the final panel version of it, the gold strip delineating the panels (I suspect by chance rather than subconscious design) can be seen as representing the Greek letter π (which is ubiquitous in astronomy theory). It also resembles a Stonehenge-style pair of Megaliths with a plinth which recalls the many astro-archaeological Megalithic stone circles in Scotland along with the extensive work by Scots (like Archie Thom, Archie Roy and Euan MacKie) on their astronomical interpretation.

The final stage in the whole process is the production of the official painting of the Armorial Bearings and writing of the certificate of approval by the Lord Lyon. This includes a purely verbal description of the image as, in the ARfs case, including two 'James Short reflecting telescopes, that on the dexter in bend sinister and that on the sinister in bend'. When asked to proof-read the text for correctness, I replied to say that surely 'that on the sinister in bend' should read 'that on the sinister in bend dexter', only to receive the very polite but wilting retort from the Lyon's Office Secretary that (of course!) anything on the sinister in bend is *always* in bend dexter unless specified otherwise!

Arms and the Man!

'Lyon here!
It's these observatory domes,
They luik like pedal bins,
They'll hae tae go!

Whit's that, 'The Lion Rampant'?
Weel… Ah mean…
We'd hae tae get permeesion aff the Queen!

An she's fair thrang wi aa this Devolution,
Can ye no fin some alternative solution?
Twa telescopes ye say?! Thon's jist the joab!

An you micht think the pillars luik lik pi,
But Heraldry's a fykie business aye,
Jist lea that tae me John… Ah'll decide!

Stonehenge ye say?! A Megalith an plinth?!
(*They tak a mile – gin e'er ye gie an inch*!)
Ah hae tae say ye've some imagination!

A constellation John? (*He's verra tryin!*)
Ye say yer favourite yin is cried' Orion'?!
Ah'll mark it in ma note buik – hae a luik!

Oh, by the way, the Saltire is OK!
An a crown – it's jist a wee yin by the way…
Nou John, aesthetic approval bides wi me!

Aince we're aa redd up we'll get tae paintin!
An naw! The costs John wullnae hae ye faintin!
A braw cerulean blue shuid dae the joab!

By the by, yer comments oan proofreadin,
Thair naethin that ye say John we'll be needin…
Aa thing's in haund! Jist lea things tae the Lyon!

Ah note ye thenk us kindly fir oor toils,
An honour tae assist Astronomer Royals!
Oh, by the way agane, keep mind the fees!'

7.4 The Scottish Dark Sky Observatory and a Burns Club Visit

Scottish Public Observatories

Close above Dalmellington in East Ayrshire, not far from Burns country, lies the world's first full-time public observatory (Scottish Dark Sky Observatory, SDSO) in an area designated in 2009 by the International Dark Sky Association as Gold Standard dark skies – Galloway Forest Dark Sky Park. Britain has four public observatories – all in Scotland – Coats Paisley (opened in 1883), Airdrie (1896), Mills Dundee (1935), and SDSO (2012). In addition, Scotland now has an IDA designated Dark Sky Island (Coll), a second Gold Standard Park (Tomintoul and Glenlivet 2018)) – the world's most northerly – and many more modest dark sky viewpoints. These may appear to be an indulgence to please star-gazing geeks, so here, as well as conveying the joy of a visit to one, we first address the wider social and ecological impacts of wasting energy in general and of light pollution in particular.

Energy Wastage and Light Pollution

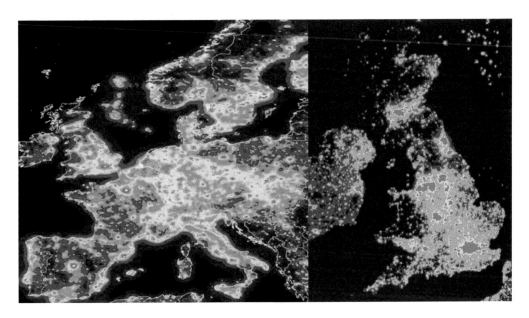

Figure 7.4.1 Europe and UK light pollution
(www.darkskyscotland.org.uk/darksky.html)

In times not long past, the area around SDSO was a major centre for coal-mining as an energy source – now universally recognised by all but some deranged politicians as a dangerous source of pollution, lung disease, and global warming due to greenhouse effect of carbon burning emissions. It is then apt that SDSO in the Craigengillan Estate and the nearby town of Dalmellington should have become part of a modern focus to battle against pollutants of other kinds, including unnecessary light, to restore the area post-industrially to its natural beauty, promoting eco-friendly and dark sky tourism.

Apart from a few world leaders only interested in making a quick buck, most to-day broadly accept that generating energy and manufacturing goods by dirty means is proving increasingly very bad for the Earth, our home, involving physical and toxic chemical waste. Less well-appreciated is how much of the energy we generate is simply wasted, due, for example, to lack of heat insulation or of care in directing energy only to where it is needed, and a large component of this is unnecessary and uninformed use of lighting.

It is widely assumed that more light is a *good* thing, for reasons of health, safety and security when, in fact, the opposite is commonly the case, as is well-documented by the International Dark-Sky Association.

- Lighting empty buildings like schools, offices, factories, and medical centres at night in fact aids rather than hinders vandalism and theft.
- Ill-placed intense security lights can dazzle victims and aid attackers.
- Lighting up the night damages the habitats of many nocturnal creatures from insects to bats, birds and even mammals like foxes and deer, many of which are vital to the overall ecosystem balance.
- The large mammals include us, and it is now recognised that loss of or changes to the diurnal light/dark cycle can be damaging to human physical and mental health, especially when it involves an increase in the blue component of lighting as is common with LEDs.
- Humans need light directed downward, not laterally to dazzle us nor upwards into the night sky, ruining our view of it, both involving a huge waste of energy and hence money. Reducing power consumption by a mere 100 W for one year saves 876 KWH of energy or about £100 at current prices. So, if all of us in Scotland did only that, then Scotland alone would annually save roughly £530 million, the UK would save about £6.5 billion (near the net cost of our EU membership), the EU would save about €86 billion and the world about €0.74 trillion. For the same amount of actual light, these costs can be cut by factors of almost five (including bulb purchase costs) by using LED and other

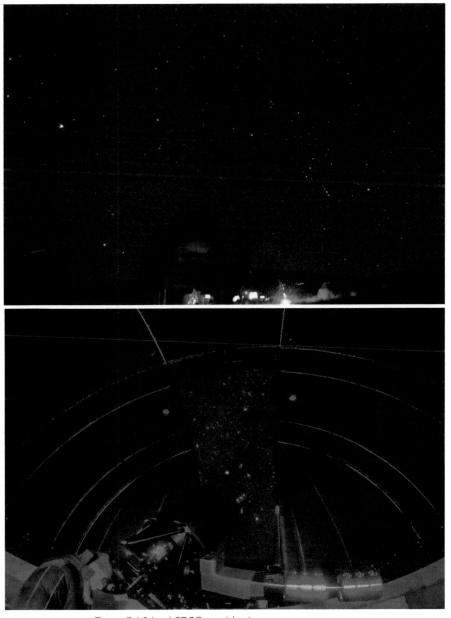

Figure 7.4.2 (top) SDSO outside view (Marcus Charron and SDSO)
Figure 7.4.3 (bottom) SDSO looking out (Marcus Charron and SDSO)

modern lamps instead of incandescent bulbs. Sadly, there is a current trend in some sectors not to reduce costs but to use LEDs to create even more glare and light pollution. However, a good example is being set by some places, including Glasgow, Skye and Dalmellington, in their road lighting replacements, which strive toward LEDs to save costs and full cut-off down-lighters to cut pollution. SDSO is a beneficiary of this policy, which is essential if the whole of Galloway Forest Dark Sky Park is to retain its status.

A Burns Club Visit to SDSO

My first night-time visit to SDSO was on 28 October 2016 as a guest of Rab's on an outing of a wee group from the New Cumnock Burns Club – a fine bunch o' Jock Tamson's bairns as I further discovered at their huge Burns Supper gathering on 27 January 2017. The day of our SDSO visit dawned wet and cloudy, typical of October in southwest Scotland, and continued thus as we prepared ourselves for the cold with pies, tattie scones and a dram at the Glenafton Athletic junior football club. However, our fortunes changed, and the skies cleared as we ascended the rocky road to SDSO, where a grand nicht was had by a' thanks to our hosts Dr Nick Martin of SDSO Trust and Ayr Astronomical Society, and Danny Cameron, amateur astrophotographer and SDSO volunteer.

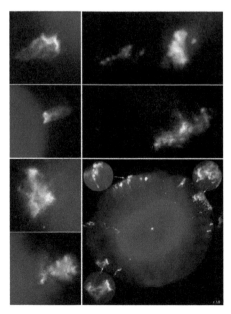

Among the many cosmic sights that we enjoyed, besides a hint of Northern Lights, were a superb naked eye view of the luminous band of the Milky Way stretching right across the sky – and a telescope view of the large Andromeda Spiral Galaxy (Messier Catalogue M31) and its two companion dwarf galaxies (M32, M110). M31 comprises about 100 billion stars plus lots of dust in spiral arms winding out from its centre. At 2.5 million LY, M31 is the farthest object visible to the naked eye and its appearance is rather like the Milky

Figure 7.4.4 Blue Snowball Nebula NGC 7662 in Andromeda with ejected 'flyers'
(B Balick, U. Washington et al., NASA, HST, WFPC2
apod.nasa.gov/apod/ap961122.html)

Way would look from there. We see the Milky Way Galaxy as a luminous band around the sky because we are inside it looking out. For more about M31 and galaxies in general see Chapter 4.

The constellation of Andromeda is also home to a lovely wee object I had never before seen or heard of before my SDSO visit – the Blue Snowball planetary nebula, NGC 7662, a class of diverse and beautiful, often complex objects discussed in Section 3.4. (PNs are all gas clouds left over from the intense mass loss in stellar winds of huge cool Red Giant stars such as the Sun will become in a few billion years.) The Snowball is blue because it is illuminated by a still very hot young White Dwarf central star.

Another PN we saw from SDSO was the classic simple ring-shaped Ring Nebula M57 in Lyra, which looks quite like the Helix Nebula shown in Figure 3.4.5(v). At the end of this section is a poem called 'Ten Draps ae Watter', inspired by seeing the Blue Snowball but also recalling what we said in Chapter 1 about the vast number of stars out there being less than atoms in our water glass.

Toward the end of our SDSO visit, there was much excitement among some of the NC Burns Club visitors when they were shown the massive Red Giant star, Betelgeuse, at Orion's right shoulder (on our left) and told that, in cosmic terms, it is getting very close to running out of fusion energy, its core collapsing and its outer layers exploding in a colossal supernova (Section 3.4). Some were disappointed that the colourful flickering they could see was just twinkling due to warm air rising from Earth and not the onset of Betelgeuse's death spectacle. When this does occur, it will outshine everything else in the sky, apart from the Sun and the Moon and be visible in daytime for weeks. Betelgeuse is far enough away (around 600 LY) that the high energy radiation from its explosion will not be a threat to Earth or to life on it. However, it is close enough (250 times closer then SN 1987a in the SMC – Section 3.4) to deliver very strong fluxes of neutrinos, gravitational waves, and photons of all electromagnetic wavelengths (62.5 thousand times those from SN 1987a), which will provide a huge wealth of data to astronomers.

Ayont the Sun

Loch Doon's silvered face reflecks the mune,
Its banks an braes hae tint her hirsel's herds,
An anely maps nou kythe firgotten wirds,
O airts whaur bairns aince dooked wi'oot their shune.
Ane bi ane the blinterin cot-hoose lichts
Hae aa gaen oot, thair anely lown stane wa's
Tae mind o thaim wha've aa bin redd awa,
Frae Starr, Black Craig, Craigmalloch an the like.
The how-dumb-deid o daurkness nou descends,
Slypes ower the Carrick Hills lik thunner cloods,
It sclimbs the ramparts whaur the Bruce aince stuid,
An aa the heichs an dowie houms it hains.
When aathing's lowsed fir aye in pit mirk nicht,
Whaur wull we cast oor een tae fin the licht?

Abune a hill that's nearhaund Craigengillan,
A metal dome revolves atop a slope,
The 20" Dall Kirkham telescope
Trains its ee oan some hyne-awa shieling;
Andromeda's antrin bricht Ring Nebulae,
Glents twa million mair's licht years awa,
The Milky Way streams oot abune us aa,
A supernova pents a starn's daith cry…
Wha wid hae thocht this laundscape purged o licht,
Wid bring sic ferlies tae wir ain hairthstane,
An braith a life intil't whaur thair wis nane.
The Cosmos birls around us glentin bricht,
An weans wha dabbled here in ilka burn,
Their progeny micht sail ayont the Sun.

Ten Draps o Watter

Inspired by the Blue Snowball Nebula

In a heivinly neuk faur, faur awa,
The Blue Snowball bides;
Wan licht year in size.

Ae drap o blue in the lift's black ocean,
Soomin amang the endless starns,
O sib Andromeda's galaxy.

The purest cerulean diamond,
Ye aamaist cuid rax an pluck frae its bield;
But aathing's no quite whit ye ettle...

Composed hail-heidit
O intangible gasses;
A cosmic will o' the wisp!

Intimatin the doulsome daith,
O a Sun jist like oor ain;
Ae star less i' the cosmos.

Ae drap o saund,
Tyned frae the hour-gless,
Blawn awa upon the solar wuins.

Ae drap o saund,
Frae aa the saund in the warld;
A puckle stour conter the firmament.

Yet aa the starns that evir wis,
Can ne'er outding the pinks o life,
Athin ten draps o watter.

Note:
'There are 120 times more water molecules in a cubic inch of water than there are stars in the observable Universe!'
(www.grc.nasa.gov/www/k-12/Numbers/Math/Mathematical_Thinking/stars_and_drops.htm)

7.5 The Kirkhill Pillar, Calderwood, and Gazing Toward Black Holes

In 1776, David Stewart Erskine – the rather eccentric but highly driven 11th Earl of Buchan – commissioned construction on his land of a model representation of the Solar System on a scale of 12,283.23 miles to the inch. All trace of it has since vanished despite, or perhaps because of, its very considerable metal content. He also commissioned the Kirkhill Pillar, a tall stone post topped with a small belfry and an iron cross in the grounds of Kirkhill House, though it was later moved to its modern location in nearby Almondell and Calderwood Country Park. It bears engravings (now much eroded) with various items including a table of predicted positions for 20 May 2255 of all major Solar System bodies known in 1776.

The reason for that choice of date is not known but it is close to 9 June 2255, the date of a rare transit of Venus across the face of the Sun. Since Buchan had a special interest in this phenomenon, specifically those of 1761 and 1769, it seems plausible that the 2255 choice of date of 20 May was related to the 9 June transit, though Buchan-era astronomer predictions should have been far better than that three-week discrepancy. (There is some unconfirmed hearsay that Buchan's tomb bears a different date in line with that of the transit.) Another factor which muddies the waters is that 2255 was at one time named the Year of the Comet because of erroneous early predictions made of a major comet returning that year.

Beginning in 2002, philanthropist Peter Stirling of Stirling Developments, Edinburgh, spearheaded the design and creation of a rather special housing and community development, Calderwood, in West Lothian, close to Almondell Country Park. This is now in existence but not yet complete. Peter's vision was of a quality eco-friendly sustainable village, but one which reflects some historic features of the vicinity, including Buchan's interests in astronomy and also old drove roads and some alleged ley-lines. One current intention is to include a small central cosmic park with astronomical features, as discussed by Peter and me, like markers for the directions of solstice sunrise and sunset, and a central black hole viewing archway. Through the latter, one could look from a marked viewing position on any given date during a few-hour period and know that one is looking toward the supermassive (100 million Suns) black hole in the centre of our Milky Way Galaxy (Chapter 4). One cannot *see* it (or any black hole) but in fact one can barely see that part of the Milky Way itself from Scotland with the naked eye. This is because it is faint and diffuse and lies so far south that it only reaches a maximum elevation of less than six degrees above our horizon. There, starlight is heavily attenuated because it has to pass through so much air, and also subjected to any local man-made light pollution. In addition, for half the year, the period when the Galactic Centre is above the horizon lies in daytime. However, simply to look at a patch of sky and just to *know* there is a massive black hole there right now excites anyone with an imagination. The fact that the light from that Milky Way area was emitted around 25,000 years ago, long before Callanish, Stonehenge and the Pyramids even existed, but well after the earliest cave paintings (Section 6.1), makes it all the more awesome. This black hole viewing idea contributed to Gill Russell's 2009–10 *Long Wave* installation in Skye's Armadale Castle Gardens, using a horn to define the black hole viewing region of the sky. In creating it she had the further inspiration that the horn is both a transmitter, as it involves us in actively looking back in(to) time and seeing information from a remote site.

If you want to have this Galactic Centre black hole experience for yourself, note that, across the UK, it is very low in the sky and near south for a few hours around midnight (GMT) in early September, and two hours earlier than that for each successive complete month – for example mid-evening in late October. Another supermassive black hole location is in the centre of M31 the Andromeda Galaxy discussed in Section 4.1. M31 is just visible to the unaided eye but is high in our night skies for a good part of the year and easily found with the aid of any online star chart, especially with the help of even small binoculars.

Finally, another black hole awareness experience is to look directly toward where we know that there is a stellar black hole. The best option is Cygnus X-1 (Section 3.4), now known to be a 15 solar-mass black hole orbiting a blue giant star of 19 solar masses. It was one of the first cosmic X-ray sources ever detected and at about 6,000 LY from us is the second-closest confirmed black hole. In order to view the location of Cyg X-1, all you need do is locate the Northern Cross pattern within the constellation of Cygnus the Swan. This is visible all night and all year (although most easily when high in the south around mid-evening in September) and lies close to the faint naked eye star η *Cygni* located about halfway along the long arm of the Cross (neck of the Swan).

CHAPTER 8
In the End

A caundle gutters,
In a winnock, hyne awa;
Syne it maun gae oot.

8.1 The End of Life on Earth and of our Planets

i) Short Term

IN CONSIDERING HERE the question of what will happen in the end to oor big braw cosmos, we will mainly address its very long-term fate in terms of global cosmic physics. This is in contrast with the relatively much shorter-term ends of life as we know it either by human follies like nuclear war, climate change, creation of drug-resistant epidemics, or hostile artificial intelligence or robots; or by random cosmic events such as giant asteroid impacts, radiation from a nearby supernova, or hostile alien attack.

ii) Medium to Long Term

As far as life on Earth is concerned, it seems certain that no form of life as we know it will survive the onset, in around 5 billion years, of the Red Giant phase of our Sun's nuclear burning evolution (see Sections 3.1 and 3.4). At that stage, the Sun will expand by a factor of 100 or more, bringing its surface close to, or possibly engulfing, the Earth and heating it to several thousand degrees, ample to melt most materials and destroy their bio-capabilities. (Here we are being conservative in ignoring forms of life unfamiliar and even unimaginable to us, except to the most creative astro-biologists and SciFi writers, examples of which were mentioned in Section 2.6.) Even by around 1 billion years, the evolving Sun will be heating our planet beyond habitability.

However, as regards life as we know it, events much earlier than the Sun's becoming a Red Giant may bring most life here to a close. In particular, the predicted slow-down and cessation of tectonic plate movement on timescales of over 100 million years, as the Earth's interior cools and becomes more viscous, will bring to an end some conditions for Earth to support life. These include recirculation from rocks back into the sea of elements vital to maintaining biodiversity in the food chain and, even more importantly, returning *enough* carbon dioxide to the atmosphere to prevent the planet from becoming extremely cold and freezing over by runaway Inverse Greenhouse Effect. In this, cooling leads to increased snowfall globally which reflects away more sunlight and accelerates the cooling. Better known publicly is the direct Greenhouse Effect in which the release of carbon dioxide, such as by fossil fuel burning, blankets and warms the Earth. This causes accelerated and eventually runaway Greenhouse Effect by permafrost thawing and release of methane, another greenhouse gas. The very high surface temperature on Venus is the result of such runaway Greenhouse Effect arising naturally, in that case due to Venus being nearer the Sun than the Earth is.

8.2 The Fates of Galaxies, Stars and Exoplanetary Systems

As discussed in Sections 1.3 and 3.4, galaxies and stars form by self-gravitational accumulation of masses of gas and dust. As such masses contract, they heat up and start to glow, powered by loss of gravitational energy. Individual masses above about 1 per cent of the Sun's become stars (rather than planets), since they get hot enough for nuclear fusion to start and keep them hot for long periods. These are longest for the lowest masses – around 10 billion years for the Sun's mass and about 100 times the present age of the Universe for the lowest mass stars. So ultimately, about 10^{12} years hence, stars and galaxies will fade into darkness, making life as we know it impossible anywhere. Almost all of life known to us, apart from simple forms using geothermal energy, works by taking in the energy of sunlight either directly, as in plants, or indirectly, as in animals, by eating plants and creatures. That solar energy intake from the hot 5,800 K Sun is utilised to create complex (low entropy) living things from (high entropy) coarse raw materials and to power their motion and other life processes. The waste heat generated by these processes is lost from organisms back into the environment and ultimately radiated away into the dark sky of our cool 2.7 K Universe. Once the stars die and cool off, that life-empowering heat-engine operating across the temperature difference between stars and empty space can no longer function.

8.3 The Final Physical Fate of the Cosmos

Since the establishment by Slipher and others around 1915–30 of the huge cosmic distance scale between galaxies and the fact that on large scales the cosmos is expanding (see Chapter 4), the great questions in cosmology have been the origin and future of that expansion. In the mid-1900s, the favoured theory was one of a Steady State Model (SSM), in which the average density of the Universe does not vary in either space or time, due to a conjectured process, linked to the expansion, of continuous 'creation' (appearance) of new matter, with no beginning and no end. However, accumulating observational evidence, including the existence and properties of the CMBR and of the higher density (extragalactic source counts) at greater distances (earlier times) increasingly favoured the Big Bang Model (BBM) – see Section 1.3 – over the SSM.

Following the Big Bang singularity, the cosmic expansion rate is governed by its initial value and by the forces acting on the matter. These include the inward-directed gravity of cosmic matter tending to slow the expansion, and any outward-directed ones like gas pressure tending to accelerate it. If the cosmic density is high enough for the net sum of these forces to be negative (in other words, if gravity wins), the cosmic expansion should ultimately stop and reverse, the matter falling back like a juggler's ball. In such a closed Universe scenario, the compression and heating of the in-falling matter might ultimately lead to a new (Big Crunch) superhot super-dense singularity at which the matter might rebound (Big Bounce) as a new Big Bang, and so on – a Cyclic Universe. This idea has philosophical appeal since the Universe has neither an ultimate beginning nor an end, and does not need to invoke creation either at one time (BBT) or continuously (SST). However, the reality is that data since the 1990s seems to demand the existence of a hitherto unknown pressure-like force which becomes dominant at very large distances and which causes the universal expansion to accelerate with time and prevent any chance of re-collapse or a cyclic cosmos. This outward force is akin to the Cosmological Constant (Λ) term which Einstein added to his General Relativistic Equations of Cosmology as an ad hoc way of preventing the cosmos collapsing under self-gravity. Einstein did not know then that the Universe was expanding, and he later referred to the introduction of Λ as his greatest ever blunder, though it has since returned as a crucial term.

At present, no one has a clue what this force can be – certainly not gas pressure – but it is given the mysterious name dark energy, some kind of property of space-time, exotic particles or some such which give space a property like a compressed sponge. This (initially) very gradual outward late acceleration by dark energy has similarities to the ultra-fast outward acceleration by inflation just after the Big Bang (Section 1.3). Whether or not this weird notion, and the evidence for acceleration, will stand the test of time remains to be seen. If it does, then the Universe will go on and on expanding and cooling far beyond when life, stars and galaxies have faded out.

This Big Freeze scenario is currently thought the most likely fate of our cosmos in the medium to long term. What the make-up and very very long-term evolution of that cold dark tenuous Universe will be is the subject of a great deal of debate and imaginative speculation. This involves inter-alia ideas from quantum mechanics, gravity, particle physics, and their implications for space and time, a bit like Big Bang theory except at the opposite physical extremes (huge instead of tiny distance and time scales; tiny instead of huge mean density and temperature, apart from at locations of black hole singularities. Even at present, the mean density of the Universe is about 10^{-30} (one thousandth of a billionth of a billionth of a billionth) times that of water.

On the utterly huge timescales discussed in some of the wilder ideas below, the Universe would be so large that there will only be about one particle in a volume around that of the whole Universe at present. To think that our knowledge of physics from observations of present cosmic conditions and from laboratory experiments can be extrapolated to conditions like these and make any sense is rather arrogant. However, it could be argued that we have no other option and that such outrageous speculative extrapolation *is* fun. So here is a small sample for those keen on such things.

Hawking Radiation from Black Holes

According to Stephen Hawking and some others, quantum effects mean that black holes cannot be perfectly black because the Uncertainty Principle means it is not possible to say with certainty whether a particle or photon is inside or outside the black hole boundary. Consequently, black holes lose mass by Hawking Evaporation though only very slowly for large masses, taking 10^{67} years for the mass of the Sun but only secs or mins for black holes of masses similar to everyday objects. They can thus in principle be seen via their Hawking Radiation. However, in the dark Universe of the distant future, the low mass bright Hawking-luminous black holes will be long gone and the slowly evaporating massive ones shining very feebly indeed. It is not certain whether the theory applies to such low masses, and it is unknown how low the masses can be of small primordial black holes formed early in the Big Bang.

Fundamental Particle Decay

In many versions of the theory of fundamental particles, some of those we think of as permanent in our everyday world are predicted to be ultimately unstable. For example, though protons can form by unstable decay of neutrons, they themselves are predicted to ultimately decay into lower mass particles, though on huge timescales of around 10^{35} years.

Nature of the Vacuum, Dark Energy and Cosmic Uncertainty

We tend to think of the vacuum of space as empty, simply because it has no visible or tangible matter in it. However, concepts like the cosmological constant Λ in Einstein's theory and today's dark energy can be thought of as forms of vacuum energy. Before, we talked about dark energy as if it was simply the source of a quite steady outward force like that from a compressed sponge, causing the cosmic expansion to accelerate. However, we really haven't a clue about the properties of dark energy which could potentially have huge implications for the future of the cosmos. The super-fast cosmic inflationary expansion invoked very early in Big Bang theory (Section 1.3) can be thought of as resulting from a blast of another form of dark energy much shorter and sharper than exists today. In other words, the dark energy force is highly time-dependent. We have no idea if it might suddenly become huge again with devastating effects implying a great deal of cosmic uncertainty in our possible futures, including:

The Big Rip
This arises in a scenario called 'phantom dark energy', where the future density of dark energy and the cosmic acceleration increase, leading to an arbitrarily high rate of space-time stretching. This ultimately causes disintegration into elementary particles of all material objects progressing downward from galactic down to the tiniest scales, the cosmos ending (as it began) in a singular state, this time with infinite dark energy density and expansion rate.

False Vacuum Catastrophes
Looking out into the Universe across space and time, one has the impression that it is stable, but it may in fact merely be long-lived. One can think of the force on cosmic matter in the force field of vacuum energy as like gravity pulling on a rock on a hillside. If the rock is at the very bottom of the hill, it is at a truly stable minimum energy, and there can be no force to make it fall further. If, however, the rock is sitting in a small hollow up on the hillside, it is in a site of local (or false) minimum energy and only as long-lived as the time until it gets a small kick out of that hollow. Depending on the shape and height of the slope below it, if dislodged, it can then plummet to a much lower energy state.

In the case of matter experiencing an analogous fall in a vacuum energy field, it may encounter very different cosmic properties and even different values of physical 'constants'. If so, the whole character of matter, space, and time in the cosmos would

Figure 8.3.1 Galaxy Puzzle
Lynette Cook: acrylic, gouache, and coloured pencil on illustration board, 1996, of which Lynette wrote:
'First created to accompany an article about the search for dark matter, this artwork pertains equally to
additional cosmic questions for which we have only partial answers – like the Big Rip idea'
(Lynette Cook)

change. Some theories of a Higgs boson-like particle allow unpredictable false vacuum collapse with near instantaneous destruction of all cosmic order! Indeed, at our present level of understanding of the laws of physics, we cannot rule out such a sudden immense change at any moment in oor big braw cosmos.

So, to all our readers, *carpe diem*. To end on a brighter note than Big Freezes, Crunches and chaos, we note that the Steady State vision of cosmology is not dead if one believes in the idea of the Cosmic Multiverse. In some versions of that, new Universes like ours are created in Big Bang white holes, connected by worm holes to black holes forming in our and other Universes, which make up an eternal Steady State Multiverse. This could mean that, at all times prior to and after the period when our cosmos is on the scene, there should be beings elsewhere in the Multiverse, enjoying the science, beauty and poetry of their own braw cosmoses.

Multiverse

Ceci n'est pas Stonehenge, this is the cosmos,
distilled to elemental rock and stone,
depicting that interstellar collision,
four billion years away, a chaos
of realignment unimaginable,
when all the worlds we knew or didn't know
osmotically pass through each other like ghosts,
to form new galaxies intangible.
Proxima Centauri, our 'close' neighbour,
the merest blink, four light years distant, seen
through Hubble may as well be some strange dream;
A single note from some symphonic score,
composed by some great sphinx-like misanthrope,
stretching far beyond all mortal scope.

Black holes, supermassive in their scale,
might tear our future Earth from that serene
orbit she's held; jolt her form terrene
to some alternate Universe far away;
after our short human race is run.
Whilst in immeasurable luminosity,
Quasars will trace out our history;
Four trillion times brighter than the Sun.
Once our brief candle long has been extinguished,
and every poem's lost and every song,
all trace of our existence lost and gone;
Except in cosmic time, there undiminished,
we'll be as shadows, alien avatars,
when we again are but the dust of stars.

Epilogue: Aeternum

Whaur aathing ends aathing begins,
Oor Universe expands, contracts;
A palimpsest rewarked in aeternum.
Oor fate hanked up in laws o physics;
Daurk energy's unkenned conflummix,
Else, aathing cuilin til a frozen waste...

Or mibbes ither warlds are dern awa,
Mangst the omnigaddrum o the spheres,
Thon eldritch multifarious multiverse,
Whaur life o sorts wull aye gang oan fir aye.
'There's howp, but no fir us', so Kafka said –
But howp there is; somewhaur ayont the starns.

Appendix

The Wider Worlds of Cosmic Imagery and Poetry

IN THIS BOOK, we have tried through our combined expertise to convey some of the magnificence of our Universe in terms of the sheer beauty of how it looks, of how scientists perceive its workings, and of the creative insights and visions it inspires in artists and poets. Our hope is that it will act as a launch pad for science fans to venture more into the space of poetry and art and learn to share the orbits of poets and artists, and vice versa, via the numerous other resources and activities available. Here, we simply suggest some starting points for further reading in the worlds of cosmic poetry and imagery, and make a few general observations about creative work across the border between science and art, in addition to those we made in our Introduction.

With regards to further reading, on the imagery side, there are copious sources of fabulous images of and about the cosmos and all space matters – photographic and artwork, professional and amateur – samples of which appear in these pages. Good starting points are: Ron Miller's *The Art of Space* (Zenith 2014); the International Association of Astronomical Artists (www.iaaa.org/); the annual volumes of *Insight Astronomy Photographer of the Year* (Collins Press/Royal Observatory Greenwich), comprising entries to the competition of that name; and a daily look at NASA's Astronomy Picture of the Day (apod.nasa.gov/apod/astropix.html). Many of the Figure credits in this book contain links to image sources, ranging from huge organisations like NASA JPL, HST, ESA, and ESO, as well as websites of individual amateur astrophotographers and artists, and of common interest groups.

For anyone into humour as well as art and science, there is a great world out there of space-related jokes and cartoons. There is no better lifting-off place into this than the material collected by the late great Professor Colin Pillinger CBE, FRS, FRGS of *Beagle* Mars Probe fame in *Space is a Funny Place* (Barnstorm Productions, 2007) and *Mars in their Eyes* (Exhibition Catalogue, 2006). Another not to be missed space-art hoot is *Mrs Moore in Space* (2002) by Gertude Moore, mother of Patrick Moore.

For further reading of general cosmic/space poetry, we can do no better than recommend the excellent anthology *Dark Matter: Poems of Space* (Gulbenkian Foundation, 2008), compiled by Professor Dame Jocelyn Bell Burnell. This anthology contains works by some of the world's leading poets, past and present, with a fair selection of Scots poets amongst them, including Norman MacCaig, Hugh MacDiarmid and Edwin Morgan, who have carried forward that ancient tradition of great Scottish makars. Other Scottish or Scotland-connected space bards include Gerry Loose and Dr Pippa Goldschmidt, while Scots-born UK poet laureate Dame Carol Ann Duffy created a lovely moon poetry collection, *To the Moon: An Anthology of Lunar Poems* (Pan Macmillan, 2009). There are numerous other space poetry writers worldwide, some wide-ranging in topic and others highly focused (for example, scientist Simon Barraclough's *Sunspots* (Penned in the Margins, 2015). However, any venture into the world of space poetry just has to include the posthumous collection, *A Responsibility to Awe* (Oxford Poets, 2001), of works by Rebecca Elson (1960–99). Her poems teem with insights, both profound and witty, into her work as an astrophysics researcher, her life in general, and her 10-year struggle against Non-Hodgkin lymphoma.

Of the several cosmic poets mentioned above, Elson is the one who penetrated most deeply the technical depths of the cosmos but, at the same time, being very much aware of the human condition, and the limitations of thinking *too* scientifically. In 'We Astronomers', she writes:

> We astronomers are nomads,
> Merchants, circus people,
> All the Earth our tent.
> We are industrious.
> We breed enthusiasms,
> Honour our responsibility to awe.

but goes on to say:

> Sometimes, I confess,
> Starlight seems too sharp…
> … I forget to ask questions,
> And only count things.

It is of interest to look at the extent to which various other poets, especially those exhibiting generally broad vision, have addressed cosmic topics, scientifically and aesthetically in their writings. Clearly scientists Dame Jocelyn Bell Burnell and Dr Rebecca Elson represent one end of the spectrum. Among Scots born poets venturing into science or space topics, we must include Edwin Morgan, a Glasgow poet laureate and subsequently Scotland's first Makar. He took a special interest in space and time issues, alongside a wide-ranging gamut of other passions. The former included JW Dunne's *An Experiment with Time*, concerning allegedly precognitive dreams, and its connections with relativity and causality. Much of his poetry, however, was about love and very human issues, including that written for the 1995 opening of Glasgow's LGBT

Centre. Even when writing about space and time, with some technical jargon, his poems were often quirky and witty in both wording and construction. Among them are his poem, 'Off Course', set aboard a spacecraft, every line being a pair of noun-phrases, and his 'First Men on Mercury', in which there is a bizarre interchange of dialogue between an Earthling and a Mercurian.

In this context, for a book incorporating Lallans poetry, it would be inconceivable not to ask where Robert Burns lies on the spectrum of addressing cosmic topics. Burns was, by all accounts, a true 'Enlightenment' man, a voracious reader all of his life, with no subject beyond his curiosity. Likewise, he and his close friend Alexander Nasmyth, the artist who had a keen interest in engineering, must surely have talked of the numerous hot

Figure A.1 Portrait of Robert Burns, 1759–96,
by Alexander Nasmyth
(National Galleries of Scotland. Bequeathed by Colonel William Burns 1872 www.
nationalgalleries.org/art-and-artists/1962/robert-burns-1759-1796-poet

topics at that time related to cosmic objects. These included events like: the 1759 return of Halley's Comet; Lomonosov's 1761 discovery of the atmosphere of Venus; Messier's 1771 catalogue of diffuse astronomical objects; and William Herschel's 1781 discovery of the new planet Uranus, expanding the known boundaries of the Solar System for the first time in history. Scottish scientists of that time did not lag behind in this pantheon of discovery: Alexander Wilson (Section 5.2), Mary Somerville (Section 5.3), and Thomas Brisbane (Section 5.4) were among contributors to this age of scientific revelation where the 'Athens of the North' punched well above its weight, as it does today. Even the great Voltaire said: 'We look to Scotland for all our ideas of civilisation.'

Given all this, and the intense interest in astronomy shown by some other visionary poets and writers from Omar Khayyam to Edgar Allan Poe, one might expect substantial content in Burns' poetry about the contents, and human knowledge, of oor big braw cosmos. However, this is not the case, as a Burns collection survey reveals. For example: the word 'star' features in many of his poems (23 times) but solely as astrological or romantic symbols, *not* as cosmic objects. For example, 'May kinder stars upon thy fortune shine' (from 'Lament of Mary Queen of Scots'); 'And her two eyes like stars in skies' (from 'O Mally's Meek'); and 'While the star of hope she leaves him' (from 'Ae Fond Kiss'). Likewise, for the word 'Sun' (or Phoebus), which is mentioned 74 times but in lines such as: 'The rising Sun, owre Galston Muirs,/ Wi' glorious light was glintan' (from 'The Holy Fair') and 'While Phebus sank beyond Benledi' (from 'By Allan Stream'). The Ayrshire world inhabited by Burns would also often show him its fields by the light of the Moon on his nocturnal outings to visit taverns and lovers. Indeed, the word 'Moon' does make 36 appearances in his poems but, again, in everyday or emotional, rather than a cosmic context. For example, 'Where the Howlet mourns in her ivy bower; And tells the midnight Moon her care.' (from 'A Vision')

There are a couple of notable exceptions to this pattern where Burns showed some awareness and vision of cosmic scales in space: 'Tho' I was doom'd to wander on,/ Beyond the sea, beyond the Sun' (from 'O were I on Parnassus Hill'); and also of the scales in time: 'And I will luve thee still, my dear,/ Till a' the seas gang dry,/ … And the rocks melt wi' the Sun' (from 'A Red, Red Rose'). This latter example seems almost to show awareness of his contemporary geologist Hutton's revolutionary ideas about Earth's great age and (more fancifully) as foreseeing the 20th century knowledge that, as a Red Giant, the Sun will eventually engulf the Earth. More realistically, his reference to the Northern Lights ('Or like the Borealis race,/ That flit ere you can point their place') shows his awareness of one key aspect of auroral behaviour physics – flickering.

While reference to the Moon and stars as cosmic bodies may be lacking in Burns' poems, their colourful portrayal of nocturnal outings certainly inspired night imagery in others. Wonderful examples are to be found among two recent artistic renderings of the 'Tam o' Shanter' saga inspired by Burns' great epic poem of that name. Most recently (2019), Nichol Wheatley's ten superbly vivid images of Tam's adventure, starting with the demon drink and ending among warlocks and witches, were installed in Glasgow's Òran Mór, the two shown in Figure A.2 exhibiting Tam and his famous horse Meg en route by moonlight across country then to the churchyard. The earlier sequence (completed in 1996) of no less than 46 images in the renowned cycle by the late Alexander Goudie (1933–2004) of 'Tam o' Shanter' paintings, include those displayed in Figure A.3 of Tam's flight across the bridge to escape the attractive Cutty Sark witch who had pursued him from the churchyard, and are infused with ominous shadows and impenetrable darkness offset by moonlight and stars. Viewing the complete set of this huge collection is a rare treat for anyone visiting Rozelle House and the Maclaurin

Figure A.2 *Tam o'Shanter* 'Whiles crooning o'er some auld Scots sonnet; Kirk-Alloway seem'd in a bleeze' both images 1.6 m x 2.6 m wide, oil on linen, 2019
(Image © Nichol Wheatley, reproduced by his kind permission)

Figure A.3 *Tam O'Shanter* by A Goudie: 'Win the Keystone at the
Brig; She Flew at Tam wi Furious Ettel' (oil on canvas 54"x72" and
60"x60"):Tam racing the witch over the Auld Brig o' Doon, the witch
snatching Meg's tail just as he reaches safety
(A Goudie and Rozelle House/Maclaurin Gallery in Ayr)

Art Gallery in Ayr.

However, the most notable exception to Burns' being oblivious to the contents of oor braw cosmos is in a letter where planets, comets and such all appear though fully mixed in with his love-life. This is in a little-known passionate love letter (stored in the Alloway Burns Birthplace Museum) to his adoring paramour of that time, one Mrs Agnes Maclehose of Edinburgh, whom he addressed by the *nom de plume* Clarinda and wrote to her as Sylvander. At the height of his infatuation, he wrote several letters to her each day, all hand-delivered. No doubt this keeps a man much fitter than email.

In the aforementioned letter, Burns touches on his dream of their escaping to a love nest on a remote planet by taking a cosmic trip beyond the near planets of our own Solar System. The full text of Burns' very long letter, started on the night of 20 January 1788 and completed the next day, was the inspiration for Rab's own adjoining light-hearted poem, 'The Star of Hope', in which the spellings are Burns' own. The key part of the letter here was on his last half-sheet of paper:

*What a strange, mysterious faculty
is that thing called Imagination?
We have no ideas almost at all, of another world; but I have often amused myself with
visionary schemes of what happiness might be enjoyed by small alterations, alterations
that we can fully enter to, in this present state of existence—For instance; suppose you
and I just as we are at present; the same reasoning Powers, sentiments and even desires;
the same fond curiosity for knowledge and remarking observation in our minds; &*

imagine our bodies free from pain and the necessary supplies for the wants of nature, at all times and easily within our reach: imagine farther that we were set free from the laws of gravitation which binds us to this globe, and could at pleasure fly, without inconvenience, through all the yet unconjecture'd bounds of creation—what a life of bliss would we lead, in our mutual pursuit of virtue and knowledge, and our mutual enjoyment of friendship and love!—I see you laughing at my fairy fancies, and calling me a voluptuous Mahometan; but I am certain I would be a happy creature, beyond any thing we call bliss here below: nay, it would be a paradise congenial to you too—Don't you see us hand in hand, or rather my arm about your lovely waist, making our remarks on Sirius, the nearest of the fixed stars; or surveying a comet flaming inoxious by us, as we just now would mark the passing pomp of a travelling Monarch?: or, in a shady bower of Mercury or Venus, dedicating the hour to love; in mutual converse, relying honor and revelling endearment—while the most exalted strains of Poesy and Harmony

would be the ready, spontaneous language of our souls! Devotion is the favorite employment of your heart; so is it of mine: what incentives then to, and powers for, Reverence, Gratitude, Faith and Hope in all the fervours of Adoration and Praise to that being whose unsearchable Wisdom, Power and Goodness so pervaded, so inspired every Sense and Feeling! – fly this time, I dare say, you will be blessing the neglect of the maid that leaves me destitute of Paper.— Sylvander

I, Rab Wilson, now offer the final of my many Scots poem contributions to *Oor Big Braw Cosmos*, based on this letter from Sylvander to Clarinda about their lovers' space trip.

Figure A.4 Exoplanet Travel Poster for PSD J318.5-22 where night life never ends
(NASA JPL)

The Star o Hope

Rabbie Burns, in purple prose,
Wrote til Nancy Maclehose,
In howps tae aiblins get some brose –
        An butter tae!
But she wis blate an douce an chose
           Tae spile Rab's day.

An sae his Muse taen tae new heichts,
(No the type his een tae dicht!),
Tae schaw oor Nancy stellar sichts,
        Ayont the mune,
Whiles oan some distant starn they'd licht,
           He'd mak her swoon!

He'd promised hissel, late or suin,
That Nancy's stays wid cam unduin,
'Voluptuous Mahometan',
        He hud nae shame,
An that his nieves wid suin impinge,
           Aroond her wame.

Seducing her oan Sirius,
Whiles comets flamed inoxious,
Tae mark some Monarch illustrious,
        Their passing pomp,
Else Mercury, or oan Venus,
           The pair wid romp!

But hinnied wirds tae sic as those,
Lik devout Agnes Maclehose?
She sees the thorn amang the Rose,
        His 'fairy fancies',
An her a 'married woman', Losh!
           There gaes Rab's chaunces.

Her guid nem's no fir the takin,
She's nae scone o yestreen's bakin!
Did he think *she'd* juist be forsakin,
   Rank an station?!
Then pey the lawin fir merry-makin,
    Wi' *her* reputation?!

Mind, at the tail end o the coo,
When Nancy's life wis aa but throu,
Her diary spak the truth enow,
   Her ae regret,
That they ne'er met again she'd rue;
    It bruck her hairt.

An sae it gangs, sic cosmic games,
Rax fir the starns, ye'll crash in flames,
Poets lik him they'll no be tamed,
   Despite their tropes,
Whiles in the lift their luve aye glaims;
    'The Star o Hope'.

I hope that wi this and ma ither Scots poems throughout *Oor Big Braw Cosmos*, I've added my own wee 'Chuckie Stane' to the cairn of Scots poetry pertaining to the cosmos.

               Rab Wilson

And, as a final astronomy thought from me, John Brown, I recall that in Section 2.6 we mentioned Exoplanet PSD J318.5-22. This is one of a rare exoplanet type which escape their parent star and travel adrift in eternal cosmic darkness but are warm enough to dance the night away until its heat of formation ultimately radiates away – perhaps a perfect place for Rabbie's romances. (See NASA JPL Tourist Poster in Figure A.4.)

               John C Brown

Map: Some Scottish Sites of Astronomical Interest

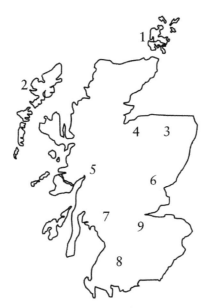

1. ORKNEY
    Maeshowe ca. 3000 BC: neolithic circular tomb; winter sunset-aligned passage
    Stones of Stennes & Ring of Brodgar ca. 2500 BC: megalithic stone circles

2. LEWIS
    Callanish ca. 2500 BC: megalithic standing stone circle complex
    Clach an Trushal ca. 2500 BC: tallest standing stone in Scotland

3. CAIRNGORM
    Tomintoul and Glenlivet Dark Sky Park: world's most northerly Gold Tier dark sky park

4. CULLODEN AND NE SCOTLAND
    Clava Cairns ca. 2000 BC: complex of cairns and standing stones; some solar alignments

5. BEN NEVIS
   Remains of weather/cloud observatory (1883–1904): inspired CTR Wilson's cloud chamber

6. DUNDEE
   Mills Observatory 1935: first purpose-built public astronomical observatory in UK

7. GLASGOW AREA
   Airdrie public observatory founded 1894: one of four in Britain, all in Scotland
   Clydespace satellite manufacturer
   University of Glasgow:
   > Bute Hall stained glass window in memory of John Pringle Nichol
   > Instrument (Lord Kelvin and Alexander Wilson) and meteorite collections
   > Inst. of Physics Blue Plaque for Lord Kelvin at No. 1 Professors' Square
   > Inst. of Physics Blue Plaque for Alexander Wilson at Acre Road Observatory
   Kelvingrove Museum: Fulton's Orrery, 1832
   Glasgow Science Centre Planetarium
   Paisley Coats Observatory, founded 1883: oldest British public observatory

8. AYR and DUMFRIES & GALLOWAY
   Arran Machrie Moor: standing stone circle
   Cumnock: Dumfries House: Grand Orrery, 1758
   Dumfries: Portrack Gardens: cosmic landscape installation art (Jencks)
   Galloway Forest Dark Sky Park, one of the world's first Gold Tier dark sky parks
   Largs: Three Sisters alignment pillars and ruins of Brisbane Observatory, 1807
   Sanquhar: Crawick Multiverse: cosmic landscape installation art (Jencks)
   Scottish Dark Sky Observatory, Dalmellington: world's first public observatory in
   Gold Tier dark sky area
   Glenlair and Parton: Childhood home and grave of James Clerk Maxwell

9. EDINBURGH AREA
   Almondell and Calderwood Country Park Visitor Centre, Kirkhill Pillar, ca. 1776
   Blackford Hill Royal Observatory Edinburgh (new ROE): Visitor Centre and famed
   Crawford Collection of historic books & instruments
   Calton Hill Observatory – Playfair Building – old ROE with transit instrument and
   Cooke telescope, now houses Collective Art Gallery; Calton Hill – One O'Clock
   Time Ball
   Castle: One O'Clock Gun, created by Charles Piazzi Smyth
   Dynamic Earth: displays of earth and planetary geology; planetarium
   Jupiter Artland: cosmic landscape artwork (Jencks)
   National Museum of Scotland: petrospheres, meteorites; historic instruments

Timeline: Some Highlights in Cosmic and Astronomical History

Note on presentation of dates:
dates over 0 are AD; negatives (-) indicate BC;
for very early times PBB = Post Big Bang; Bn yrs = billion years

-13.8x10⁹ yr	Big Bang	1659	First good measurement of Astronomical Unit (Huygens)
10⁻³³ sec PBB	Cosmic Inflation		
378,000 yrs PBB	Cosmic Recombination era	1666	Newton's Law of Gravity (& Motion, 1686)
-5 Bn yrs	Large Scale Structures start to form	1668	Newtonian reflecting telescope
-4.6 Bn yrs	Sun forms	1675	Astronomer Royal Post created
-4.5 Bn yrs	Earth forms	1776	Black hole: concept, John Michell; named by John Wheeler, 1967
-4.3 Bn yrs	Oldest fossil life		
-0.16 Bn yrs	First mammals	1781	Prediction and discovery of Uranus
-200,000	Homo Sapiens	1834	Astronomer Royal for Scotland post created
-50,000	First cosmic cave paintings		
-5200	Neolithic tombs (Maeshowe/ New Grange)	1838	First stellar distances measured – (Bessel; Henderson 1839)
-5000	Standing stone circles	1840	First astro-photography (Daguerre)
-4500	Pyramids	1845	Leviathan (72-inch) telescope, Parsonstown
-4000	Clava Cairns/petrospheres		
-200	Aristarchus heliocentric model	1846	Discovery of Neptune
100	Ptolemy's *Almagest* – Greek treatise	1856	Tenerife mountain-top observing survey (Piazzi Smyth)
700	Sunspots – oldest known mention (China)	1859	Maxwell theory of Saturn's rings
		1864	First astronomical line spectro-scopy (Huggins)
800	Islamic astronomy golden age		
1054	Chinese observe Taurus supernova (Crab Nebula)	1873	Maxwell's Equations
		1912	Cosmic expansion first detected (Slipher)
1128	Sunspots – first image, John of Worcester		Cepheid Law and distance meas-urement (Henrietta Leavitt)
1259	Nasir al-Din al-Tusi's Maragha Obervatory (Persia)		Cosmic Rays discovered (Hess)
1543	Copernicus *de revolutionibus orbium coelestium*	1915	M31 distance first measured (Opik and others)
1570	Tycho Planetary Observations		General Theory of Relativity (Einstein)
1609	Kepler's Laws		
1610	Sunspots: first telescopic observation – Thomas Harriet (pre-Galileo)	1927	Big Bang theory (Lemaitre)
		1929	Hubble law of cosmic expansion
	Galileo Telescope	1930	Discovery of Pluto

Year	Event	Year	Event
	Theory of white dwarf stars (Chandrasekhar)		Cosmic neutrinos detected
1933	Dark Matter concept (Zwicky and others)	1990	First Light Keck Telesope (twin 10 m)
	Dawn of radio astronomy (Jansky)		Hubble Space Telescope launched
1948	Steady State Theory (Bondi, Gold and Hoyle)	1995	First (non-pulsar) Exoplanet detection 52 Pegasi
	Prediction of microwave Background (Gamow *et al.*)	1998	First Light ESO Very Large Telescope (quad 8.2 m)
1949	First Light Mount Palomar Telescope (200-inch)		Cosmic Acceleration detected (Perlmutter, Schmidt, Reiss)
1953	Adaptive optics invented (astro applications ca. 1990)	2001	*Voyager 1* – First probe to leave the solar system
1957	Jodrell Bank Lovell Telescope first light	2002	Solar neutrino paradox solved
	Sky at Night programme starts	2006	Planet Pluto recategorised Dwarf Planet
	Theory of stellar nucleosynthesis (Hoyle, Fowler & Burbidges)	2010	International Space Station completed
	First Earth satellite *Sputnik 1*	2012	Opening of Scottish Dark Sky Observatory, Dalmellington
1959	First probe on the Moon – *Luna 2*	2015	First detection of gravitational waves (black hole merger)
1961	First man in orbit – Yuri Gagarin aboard *Vostok 1*		LISA *Pathfinder* mission
	First interplanetary probe – *Venera 1*	2017	Gravitational wave detection of neutron star merger
1962	Dawn of x-ray astronomy	2018	*Parker Solar Probe* launched
1964	Discovery of microwave background (Penzias and Wilson)	2019	Chinese probe *Chang'e 4* lands on lunar far side
1967	Discovery of pulsars (Jocelyn Bell Burnell)		Asteroid sampling by *Hyabusa 2* and *Osiris Rex* missions
1968	Solar neutrinos detected (Goldstake Mine)		First ever black hole image
1969	First man on moon – Neil Armstrong	2020?	ESA *Solar Orbiter*
1970	First planetary (Venus) landing probe – *Venera 7*	2021?	Launch of James Webb Space Telescope
1971	First space station – *Salyut*	2030?	Launch of LISA space gravitational wave detector
1973	*Apollo* Skylab Space Station		First human flight to Mars
1974	Binary pulsar discovered (Hulse & Taylor)	???	Discovery of fossils on another planet
	Hawking Radiation Theory	???	Discovery of life on another planet
	First Mercury probe – *Mariner 10*	???	Discovery of or signal from intelligent life on another planet
	First Jupiter Flyby – *Pioneer 10*		
1976	First Mars landing probe – *Viking 1*		
1986	NASA Shuttle *Challenger* disaster		
1987	Supernova 1987a in LMC		

Luath Press Limited

committed to publishing well written books worth reading

LUATH PRESS takes its name from Robert Burns, whose little collie Luath (*Gael.*, swift or nimble) tripped up Jean Armour at a wedding and gave him the chance to speak to the woman who was to be his wife and the abiding love of his life. Burns called one of the 'Twa Dogs' Luath after Cuchullin's hunting dog in Ossian's *Fingal*. Luath Press was established in 1981 in the heart of Burns country, and is now based a few steps up the road from Burns' first lodgings on Edinburgh's Royal Mile. Luath offers you distinctive writing with a hint of unexpected pleasures.

Most bookshops in the UK, the US, Canada, Australia, New Zealand and parts of Europe, either carry our books in stock or can order them for you. To order direct from us, please send a £sterling cheque, postal order, international money order or your credit card details (number, address of cardholder and expiry date) to us at the address below. Please add post and packing as follows: UK – £1.00 per delivery address; overseas surface mail – £2.50 per delivery address; overseas airmail – £3.50 for the first book to each delivery address, plus £1.00 for each additional book by airmail to the same address. If your order is a gift, we will happily enclose your card or message at no extra charge.

Luath Press Limited
543/2 Castlehill
The Royal Mile
Edinburgh EH1 2ND
Scotland
Telephone: +44 (0)131 225 4326 (24 hours)
email: sales@luath. co.uk
Website: www. luath.co.uk